ISBN 978-1-334-19533-4
PIBN 10614369

This book is a reproduction of an important historical work. Forgotten Books uses
state-of-the-art technology to digitally reconstruct the work, preserving the original format
whilst repairing imperfections present in the aged copy. In rare cases, an imperfection in
the original, such as a blemish or missing page, may be replicated in our edition. We do,
however, repair the vast majority of imperfections successfully; any imperfections that
remain are intentionally left to preserve the state of such historical works.

1 MONTH OF
FREE
READING

at

www.ForgottenBooks.com

———————◆———————

By purchasing this book you are eligible for one month membership to ForgottenBooks.com, giving you unlimited access to our entire collection of over 1,000,000 titles via our web site and mobile apps.

To claim your free month visit:
www.forgottenbooks.com/free614369

PRINTED BY
SPOTTISWOODE AND CO. LTD., NEW-STREET SQUARE

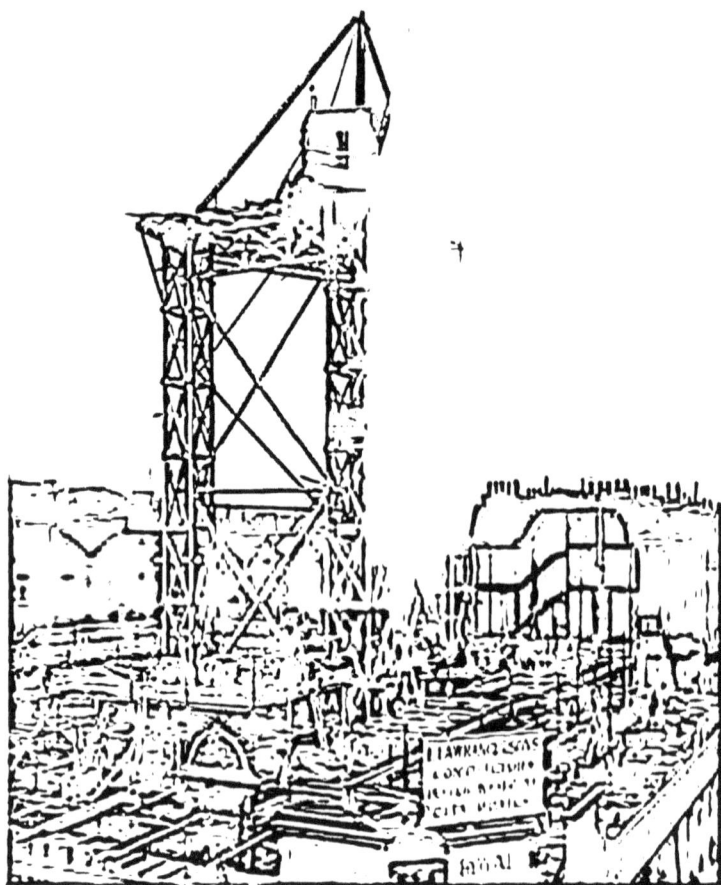

DERRICK STAGING.

SCAFFOLDING

A TREATISE ON
THE DESIGN & ERECTION OF SCAFFOLDS,
GANTRIES, AND STAGINGS,

With an Account of the Appliances used in
connection therewith

FOR THE USE OF CONTRACTORS, BUILDERS,
CLERKS OF WORKS, ETC.

With a Chapter on the Legal Aspect of the Question.

BY

A. G. H. THATCHER,
BUILDING SURVEYOR

WITH 146 DIAGRAMS AND 6 FULL-PAGE PLATES.

LONDON:
B. T. BATSFORD, 94 HIGH HOLBORN.
·1904·

CALLED

PREFACE

SCAFFOLDING up to quite recent years has been considered by builders and others concerned, with the exception of the actual workmen, to be a matter of small importance and consequently unworthy of study. Recent legislation, however (the Workmen's Compensation Act, 1897, and the Factory and Workshop Act, 1901, and preceding dates), has brought it into greater prominence, with the result that more attention has lately been given to it. Special study of the subject has, however, remained very difficult owing to the lack of accessible information.

The author, in the course of considerable experience in the building trade, has had opportunities of examining a large number of scaffolds throughout the country, affording him exceptional facilities for thoroughly studying the subject, and he has been led to prepare this treatise in the hope that it may prove useful to those engaged both in the design and erection of scaffolds for building purposes. While special attention has been given to practical details, the theory has not been neglected, but has been dealt with by the use of terms well understood in the building trade. The various formulæ given have been simplified as far as possible, and it is hoped that in these days of technical education they will not be beyond the scope of the reader.

The illustrations have generally been drawn to scale, but for the sake of clearness, details are given to a larger scale where necessary.

The practice of allowing workmen to erect scaffolds without the aid of expert supervision, as is generally the case, is to be strongly deprecated. The architect, builder, or clerk of works, should in all cases be responsible for their erection—the risk of defective or unsafe work being thereby minimised, and an economy often effected in both labour and material.

The author desires to acknowledge his indebtedness to Mr. G. Thatcher, of H.M. Office of Works and Hampton Court Palace, for valuable information contributed by him, and to Mr. J. Clark, of the Factory Department of the Home Office, for his very careful reading of the proofs ; while his best thanks are due to the following manufacturers :—Mr. C. Batchelor, Messrs. Bullivant & Co., Ltd., Messrs. Butters Bros., Mr. J. Fishburn, Messrs. Frost & Co., and Mr. E. Palmer, who have furnished him with particulars of their various specialities.

A. G. H. T.

LONDON : *February* 1904.

CONTENTS

CHAPTER V^I

CORDAGE AND KNOTS

CHAPTER VI

SCAFFOLDING ACCESSORIES AND THEIR USE

CHAPTER X

THE PREVENTION OF ACCIDENTS

CHAPTER XI

LEGAL MATTERS AFFECTING SCAFFOLDING

LIST OF ILLUSTRATIONS

LIST OF ILLUSTRATIONS

SCAFFOLDING

CHAPTER I

SCAFFOLDING

SCAFFOLDING is the art of arranging and combining pieces of timber in order to enable workmen to proceed with their work, and from which, if required, to lift and carry the material necessary for their purpose. Many definitions of a scaffold have been given by authorities on building construction ; some of the best known are as follows :—

Mitchell (C. F.) : ' Temporary erections constructed to support a number of platforms at different heights, raised for the convenience of workmen to enable them to get at their work and to raise the necessary material for the same.'

Tredgold (Hurst) : ' A scaffold as used in building is a temporary structure supporting a platform by means of which the workmen and their materials are brought within reach of their work.'

Rivington : ' Scaffolds are temporary erections of timber supporting platforms close to the work, on which the workmen stand and deposit their materials.'

Banister F. Fletcher, in ' Carpentry and Joinery ' : ' A scaffold is a temporary structure placed alongside a

B

building to facilitate its erection by supporting workmen and raising materials during the construction, or for the repair of buildings.'

Recent cases tried under the Workmen's Compensation Act have given a wider meaning to the word, and the following definition is perhaps the most comprehensive at the present time :

A scaffold, as used in building, is a temporary arrangement of timbers combined and supported in various ways to enable the workmen to proceed with their work, and where required, to afford facilities for the lifting and carrying of the materials.

The two principal methods of scaffolding are known respectively as the North and South country systems. The northern, as indicated by the name, was at one time in use only in Scotland and the north of England, but its many advantages, more especially for the transport of material, have now caused it to become general throughout the country.

The second method is essentially the South country system, and is of greater use when power is not necessary for the construction of the building.

A combination of both methods is commonly seen, and found useful in practice.

In scaffolding, the vertical timbers are known as standards or uprights. The horizontal timbers between the standards are known as ledgers when of cylindrical section, but as transoms and runners when of rectangular section. Braces, shores, struts and ties of any section are pieces used to stiffen the structure. The putlogs, or joists as they are called when of greater length, carry the boards which form the working platform.

The Northern System.—This scaffolding can be divided into two parts. First, the derrick staging from which the transporting power acts ; and, second, the platforms, which bring the workmen within reach of their work.

Derrick Stagings.—These stagings, also known as Scotch derricks and 'Scotchmen,' are erected to carry the power required, usually a steam crane.

They consist of three or four timber towers or legs supporting a platform upon which the crane stands. The number of legs depends upon the area over which the power is required to act.

When one crane is to be erected, three legs are sufficient to carry the platform.

If the building is a large one, several such stagings may be constructed ; but in some cases two cranes are required where the size of the building will not allow of two stagings. In these cases the platform is square and supported at each angle by a leg. The cranes are then fixed diametrically opposite each other.

In determining the position of the legs they must be placed where the effective range of the crane is most ·required, and also where they will cause the least possible obstruction to the progress of the building. The position of the tower that carries the crane, and which is known as the principal or king leg, is first fixed. The secondary or queen legs are set out from it in the form of an isosceles triangle. The distance between the king and queen legs depends upon the length of the sleepers. These run from below the engine to the lower ends of the guys, and average from 25 to 30 feet in length.

The legs, especially the king legs, if intended to rise from the earth, must have a foundation of two thicknesses of 3-inch timbering laid crosswise. This is unnecessary if there is a concrete or other solid foundation.

The legs in this manner can be made to support a platform up to 120 feet in height.

The required height having been reached, the legs are connected by trussed beams in the following manner : Two balk timbers of about 12 in. by 8 in. are laid immediately above each other between the king leg and each queen leg, resting on the two top transoms, as shown in fig. 1. They are from 6 to 9 feet apart, the top bay being sometimes made slightly lower than the others.

The lower balks are connected to the centre standard of the king leg by wrought-iron straps.

FIG. 2.—PLAN OF KING LEG

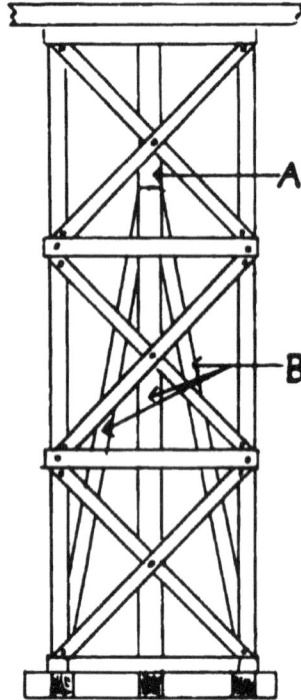

A, Central Standard. B, Shorings

FIG. 3.—SHOWING SHORING TO CENTRAL STANDARD

The top balks project from 6 to 10 feet beyond the king leg, and are halved at their point of intersection The projecting ends are connected to each other by pieces 8 in. by 6 in., and again to the return balk by similar pieces (see fig. 4). They are also supported by struts from the central standard, as shown in fig. 1.

The upper and lower balks are connected by iron bolts about 10 feet apart, and each bay thus formed is cross-braced in the same manner as the legs.

The iron bolts are covered by pieces of the same scantling as the braces.

In the single derricks the queen legs can be connected by a trussed beam similarly formed, or by a single balk carried across and laid on the top transom.

If the span is considerable, struts can be carried

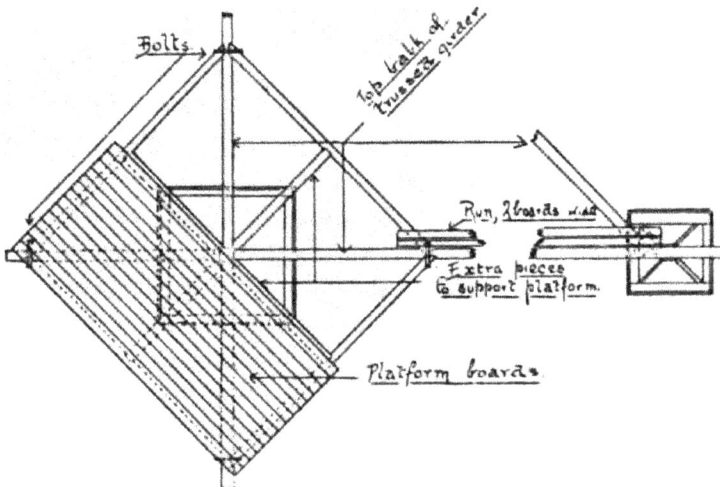

FIG. 4.—PLAN OF TOP PLATFORM PARTIALLY COVERED

from the queen legs towards the centre of the underside of the balk to prevent sagging.

On the trussed beams thus formed, joists of 9 in. by 3 in. or ordinary poles are laid about 3 feet apart.

They are laid parallel to one another, and in a direction at right angles to the truss or single beam forming the back support of the platform.

The centre joists are continued to the ends of the balks which project beyond the king leg.

The advantage of having continued the top balks

can now be seen, as it gives greater area to the platform immediately round the engine.

The boards 9 in. by 1½ in. are laid at right angles to the joists.

Another way of forming the platform is to cover only partially the surface between the legs. In this case two additional joists, 6 in. by 6 in., are thrown across the king leg (see fig. 4), the boards not extending beyond their length.

When this is done, the workmen reach the platform from the communicating ladder which usually passes up a queen leg, by means of a run two boards wide. It is better to lay the larger platform, as, apart from the question of safety to the men, it serves as a storage for coal for the engine, the weight of which tends to keep the erection steady. Double boards should be laid under coals or other heavy stores.

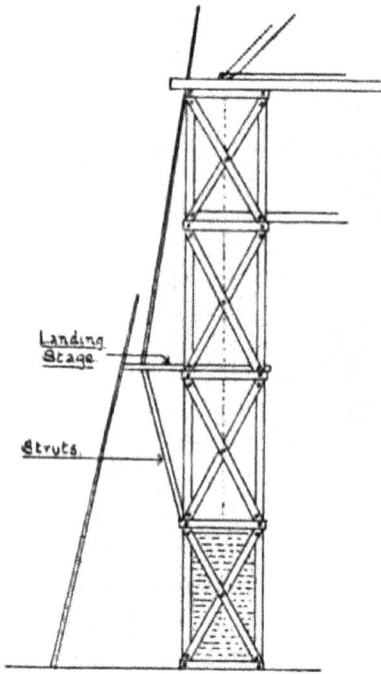

Landing
Stage

Struts

FIG. 5.—SHOWING METHOD OF FIXING LADDERS

To reach the platform, ladders are fixed in different ways. They can run up inside, or be fixed to the outside of the queen legs. In either case they are nearly or entirely upright. A better method is shown in fig. 5, and should be carried out wherever possible.

The derrick sleepers, two in number, are of balk timber, and lie across the platform from beneath the engine bed to which they are connected, to the centre of the queen legs.

The guys or stays, also of balk timbers, besides being connected to the mast, are attached to the sleepers over the queen legs (see fig. 1).

To counteract the overturning force exerted by the jib and the material lifted, the guys are chained down to the timber balk at the bottom of the queen legs (fig. 1).

This balk supports a platform which is loaded with bricks or stones more than equal to double the weight that will be lifted. The chain, which works loose with the vibration of the scaffold, is tightened by means of a screw coupler fixed in its length. The arrangement is as follows :—Two lengths of heavy chain with large links at each end are required. One length is carried round the sleeper and then taken down the centre of the leg. The other length is taken round the balk which is placed underneath the staging, and carried up through the load, when the tightening screw can be applied and the correct tension brought up.

To prevent lateral motion the legs are cross-braced by poles or deals between each leg as shown on frontispiece. The poles are tied to the legs just beneath the platform and connected at their meeting point. When crossing they should be at right angles to each other.

Deals 9 in. by 3 in. can take the place of the poles if required, bolts in this case being used instead of tyings.

At the building of the new Post Office, Leeds, 1893, a different method of raising the platform for the crane was adopted. The legs, instead of being framed, consisted of a single balk of timber strutted on each side from the ground level, the sleepers and guys being firmly attached to the standards themselves.

When erecting long ranges of buildings it may be more convenient to have the derrick mounted upon a travelling bogie than to dismantle the structure in order to re-erect at another point.

Fig. 6 illustrates the system, the travelling power being usually manual. The arrangement is suitable for small derricks, and is employed where the crane is erected outside the building.

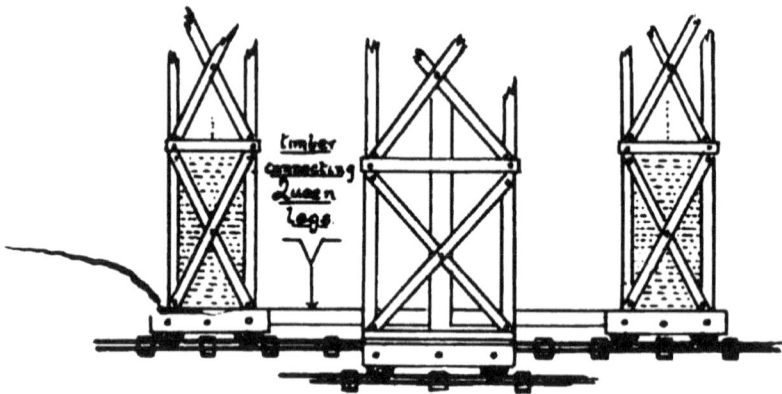

FIG. 6.—SHOWING STAGING MOUNTED ON TRAVELLING BOGIE

Another method of using travelling cranes is to erect a platform as shown in fig. 7.

The standards, which may be of balk timber or built up, as previously shown, are about 10 feet apart longitudinally and 20 to 30 feet transversely. They stand upon sills of the same section where the foundation is not solid. On the head of the standards, the runners are laid connecting all the standards in the same row.

Head pieces may be fitted between the standards and runners ; this serves to distribute pressure. All the connections are securely made by dog irons, bolts, and

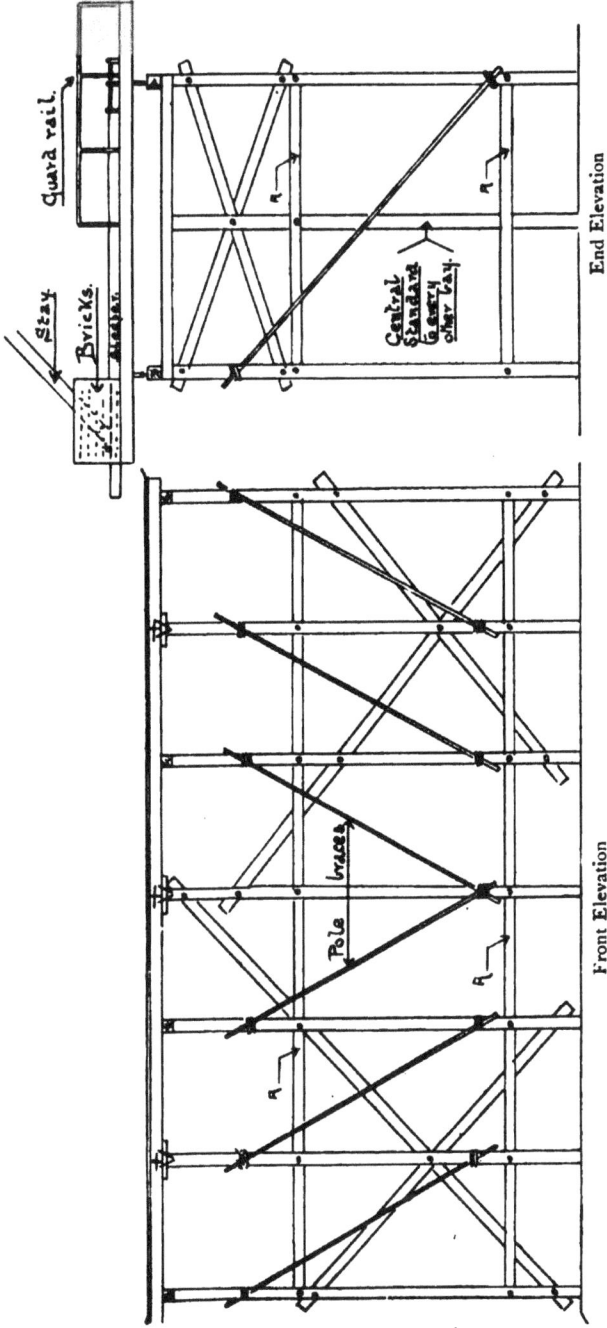

Guard rail

Stay

Bricks.

Mortar.

Central
Standard
(a every
other bay.)

R

R

End Elevation

Pole braces

R

R

Front Elevation

FIG. 7.—ELEVATION OF DERRICK STAGING

straps. The stability depends entirely upon the bracing, and this, it is important to note, should be between each bay longitudinally, and at least every second bay transversely.

Timbers placed as A in fig. 7 give rigidity to the standards by preventing flexure, and are necessary when the lengths of the uprights exceed 30 times their least diameter.

The deals used for braces are bolted to the standards ; for poles, tying is resorted to.

Working Platforms.—The working platforms used in conjunction with overhead or overhand work depend upon the requirements of the building.

By over head or hand work is meant that the material upon which the mechanic is to be employed reaches him from over head or hand.

When no outside scaffolding is needed, the platforms are laid upon the floor joints in the interior of the building, being raised upon trestles as the work proceeds, and until the next floor is reached.

Light forms of scaffolds, as the ordinary masons' and bricklayers' pole scaffolds, are now frequently used as working platforms in connection with the Scotch system.

The South Country System.—This system is divided into two classes according to the strength required. For the first, square timbers are used ; for the second, poles are employed. The scaffolds built of square timbers are known as gantries and stagings, and the pole erections are termed bricklayers', and masons' or independent scaffolds.

Gantries.—The term gantry was originally given to *erections* constructed with a view to the easy carriage of heavy material, but of late it has also come to mean a structure arranged to support lighter forms of scaffolding over footpaths which have to be kept open for public use.

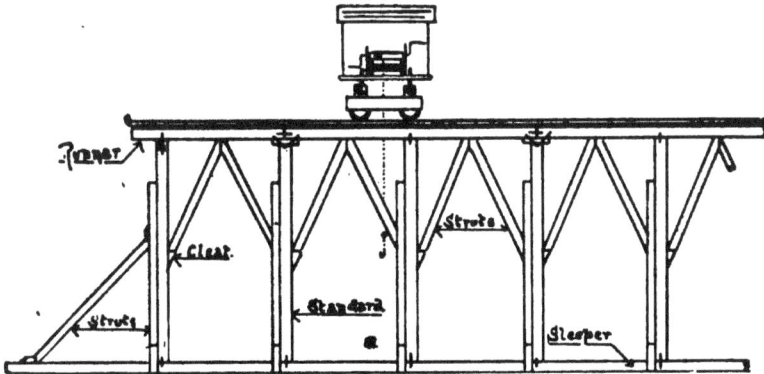

FIG. 8.—ELEVATION OF GANTRY FOR TRAVELLER

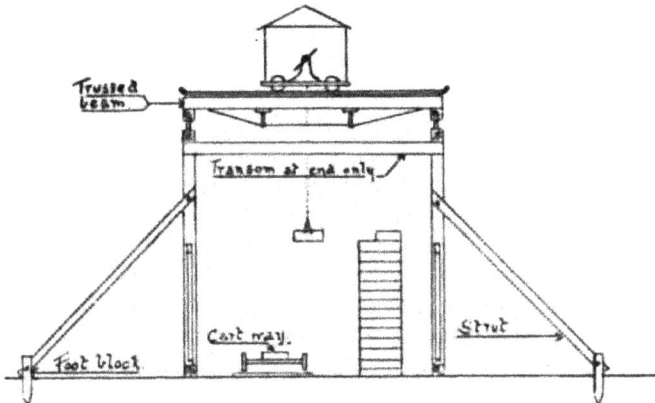

FIG. 8A.—END ELEVATION OF GANTRY SHOWN IN FIG. 8

1st. Gantries for transport of material, commonly called travellers. Figs. 8 and 8A show the general construction.

The distance between the outer rows of standards and the wall depends upon circumstances. If possible, the space should be allowed for a cart-way, as the material can thus be brought quite close to the work before being lifted. If, owing to adjacent footpaths or any other reason, this cannot be done, the uprights should be placed close to the wall on either side, the material being lifted at the end of the gantry or other convenient spot, over which the lifting gear can be brought.

The standards of square timber for the gantry are from 6 in. to 12 in. square, and are erected upon sleepers, or, as they are sometimes termed, sills laid in the same direction as the run of the scaffold. One row of standards is placed on each side of the wall. The standards are placed 8 to 10 feet apart. On the top of the standards runners are fixed connecting each standard in the same row. Sills, standards, and runners should be of the same sectional area. The runners are strutted on their underside, from the standards by pieces of, at least, half the sectional area of the supported timbers. If the struts are of equal size to the runners, double the weight can be carried.

The cleats from which the struts rise, are simply spiked to the standards, but if designed to carry excessive weights they are slightly housed in. As the space between each row of standards has to be kept open for the building, no cross bracing can be allowed except at the ends. Strutting is therefore resorted to in order to give stability. The struts, one to each standard, are bolted to the upright near the top, and again to a foot block driven into the ground. Other methods of fastening down the bottom ends of the struts are shown in fig. 9 ; the use of each depends upon the nature of the soil.

Struts are also fixed at the ends to prevent lateral movement. Head pieces, or corbels, as they are sometimes termed, are occasionally inserted between the standards and runners, and serve to distribute pressure.

Straining pieces spiked on the underside of the runners, for the struts to pitch against, are used when the standards are considerably apart.

Rails upon which the travelling engine or traveller can move are laid on top of the runners, and are turned

FIG. 9.—FOOTING BLOCKS FOR STRUTS

up at the ends of the platform to serve as buffers to the engine platform.

The engine platform consists of two trussed beams of timber about 3 feet apart, connected at their ends with short pieces of the same scantling, and fitted with

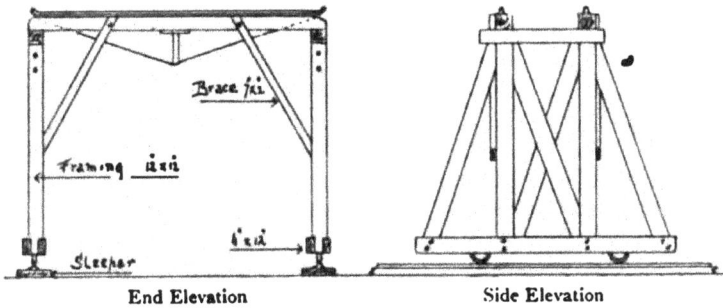

End Elevation   Side Elevation

FIG. 10.—ELEVATION OF TRAVELLING GANTRY

grooved wheels to move upon the rails. Rails are also laid upon each beam and serve for the traversing motion of the crab. Movement of the traveller is obtained from the crab, which is worked either by manual or steam power, and acts through a system of shafting and geared

wheels.   Movement in three directions is necessary
from the crab: vertically for lifting, and horizontally
in two directions, transversely and longitudinally.
Travellers are made up to 50 feet wide and any required
length.

Another method of building travellers is shown in
fig. 10.

In this case, the rails upon which the traveller moves
in a longitudinal direction are fixed on sleepers on the
ground level, and the standards and runners of the first
example are not required.   In their place is constructed
a triangulated system of balk timber framing.   The
platform is fixed to the head pieces, and is braced as
shown.   Less timber is used in their construction, but
owing to the greater weight a steam winch is required
to impart motion.

**Gantries which serve as a base for lighter forms
of scaffolding.**—These erections are in reality elevated
platforms, and allow of a clear way for a footpath where
required.   They are constructed of two frames, placed
apart according to the width of the path over which the
platform stands (fig. 11).

The method of erection, so far as the side frames are
concerned, is the same as for the first example of
travelling gantries.   Stability is, however, gained by
cross-bracing as shown in figure, thus making strutting unnecessary.   The platform can be laid by placing
short boards 9 in. by 3 in. across the runners when
the platform is narrow.   It is more usual, however,
to place joists 10 in. by 2 in. across, and on these
to lay the boards longitudinally.   The joists average
2 to 3 feet apart, the braces are about 2 in. by 7 in.
On the outside of the scaffold, parallel to the sills, balk
timbers are placed forming a ' fender ' to prevent the

vehicular traffic from injuring or disturbing the erection.

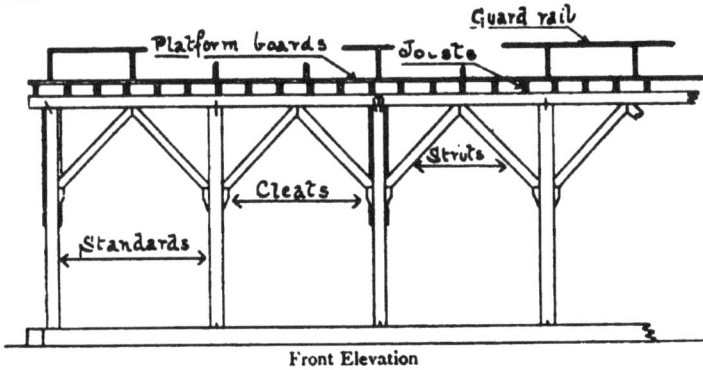

Front Elevation

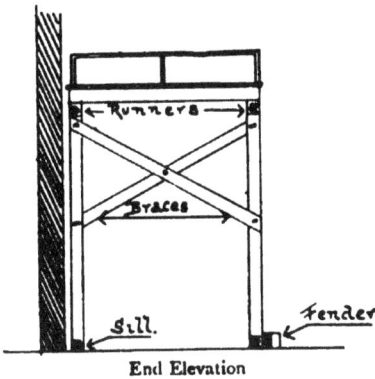

End Elevation

FIG. 11.—GANTRY OR ELEVATED PLATFORM OVER FOOTPATHS

Stagings. — Stagings are erected in a manner similar to travelling gantries, but are carried more than one storey high (fig. 12). It is a form of scaffolding rarely seen, more especially since the introduction of the Scotch derrick system. The timbers are erected to the height of the first runner in the same manner as the frames in fig. 11. In order to carry the scaffold higher, horizontal pieces are laid across the scaffold, over the standards, and are made to project 9 or 10 feet on each side of the runners.

On these beams, uprights, as in the first tier, are raised, being connected in like manner, longitudinally by transoms. The rising tiers of standards are strutted by timbers A A, rising from the projecting portion of the beam called the footing piece, which serves in the same

C

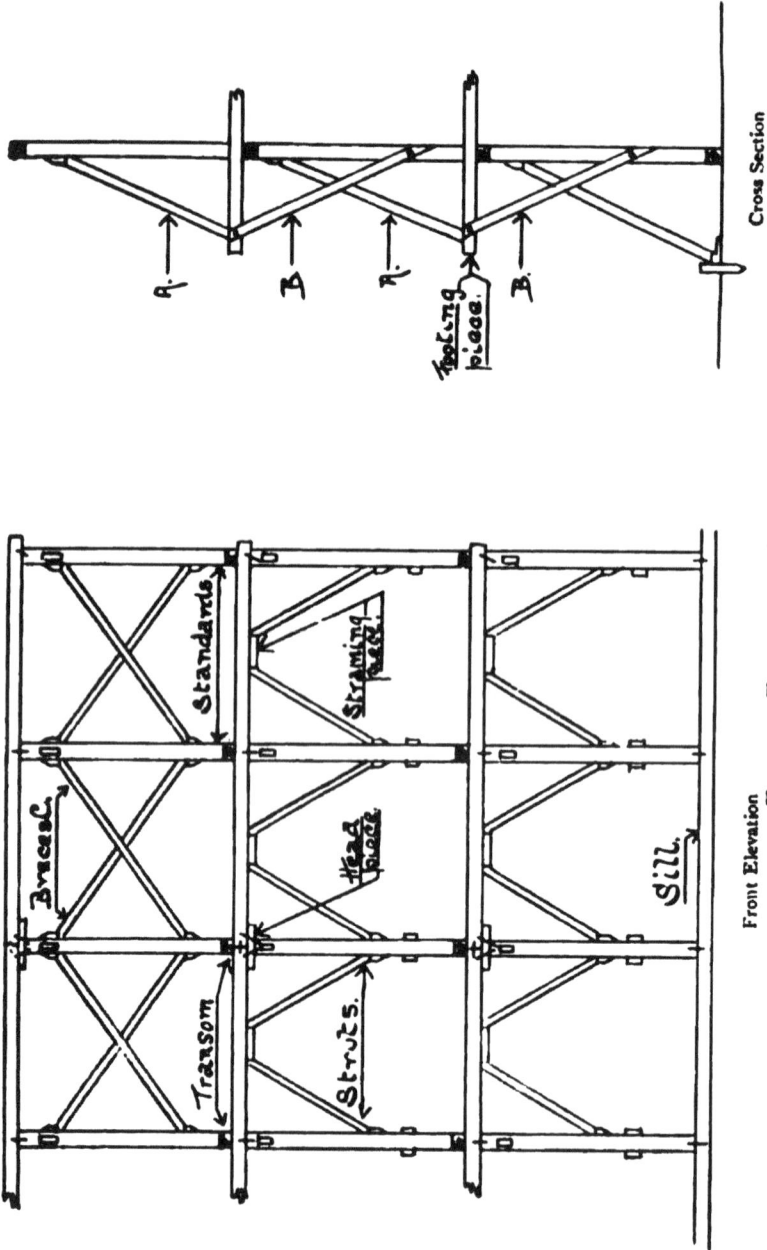

Cross Section

Footing piece.

A.

B.

R.

B.

Standards

Straining piece

Head piece

Struts.

Brace.

Transom

Sill.

Front Elevation

FIG. 12.—EXAMPLE OF STAGINGS

manner as a footing block. The footing piece is supported by struts, B B, rising from the lower standards. The struts B B are in two pieces, being bolted to the sides of the footing pieces and uprights. This allows the shores A A to pass between them.

Strutting within the bays formed by the standards is carried out on each tier with the exception of the top, where braces are fixed, as shown at C.

On the top runners rails are laid for a traveller.

In constructing the foregoing square timber erection, note should be taken of the following points :—

That the uprights of the upper tiers should stand immediately over those of the lower tiers, in order to prevent cross strains on the runners.

That the timbers should fit as evenly as possible, as thereby the whole erection is rendered more stable.

That joints between the runners should occur immediately over the standards.

The several parts of this structure, if for temporary purposes, can be connected by dog irons ; if for a more permanent use, by bolts and straps.

POLE SCAFFOLDS

Bricklayers' Scaffolds.—A bricklayer's scaffold consists of a series of upright poles or standards, to which are lashed horizontal poles, termed ledgers. The ledgers and the wall of the building carry the putlogs, on which boards are laid to support the workman, his material, and tools (fig. 13).

The standards are first erected, and may stand singly or in pairs. In a repairing job, unless of great height, and where there is no great weight of material, single poles are sufficient.

C 2

Where double poles are required, the first pair are erected of different lengths.

The short pole is termed a puncheon. .The difference of length allows of a lap in connecting the succeeding poles.

The lap should equal half of the full-length pole.

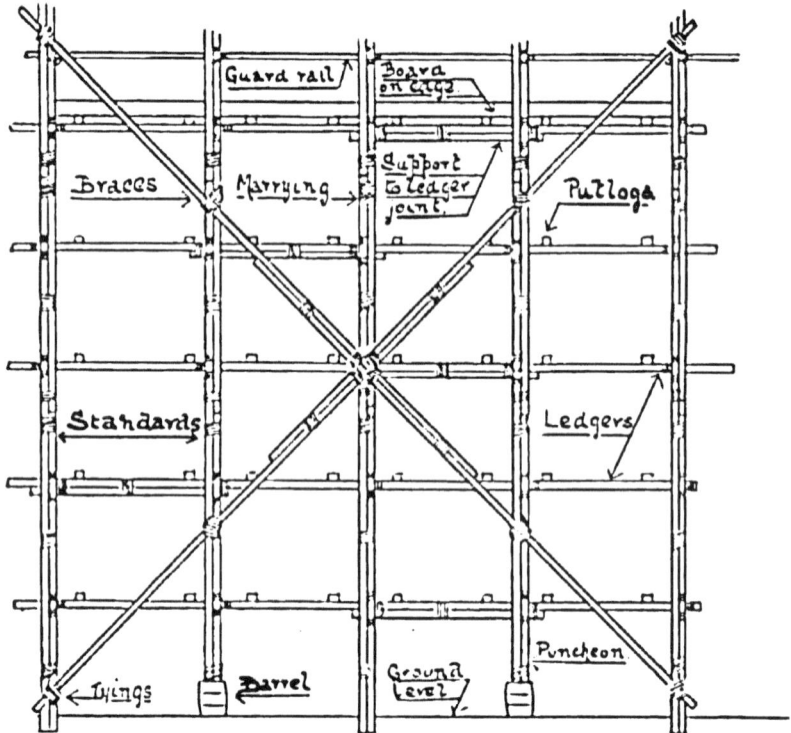

FIG. 13.—ELEVATION OF POLE SCAFFOLD

The standards are placed 6 to 8 feet apart, and from 4 to 6 feet away from the building.

The butt-ends are embedded about 2 feet in the ground, which affords some resistance to overturning. If they cannot be embedded, they should be placed on end in barrels filled with earth tightly rammed.   As the

building rises additional poles are added, being lashed
to the standards already erected.

If the standard is a single pole, the second pole,
having a lap of 10 or 15 feet, stands upon a putlog
placed close to the first pole for that purpose (fig. 14).

The inner end of the putlog is securely fastened
down to the scaffold or inserted into the building.

If the standard is double, the rising pole is placed
upon the top end of the puncheon, and afterwards
others are placed on end upon the lowest free end of the
standards already fixed.

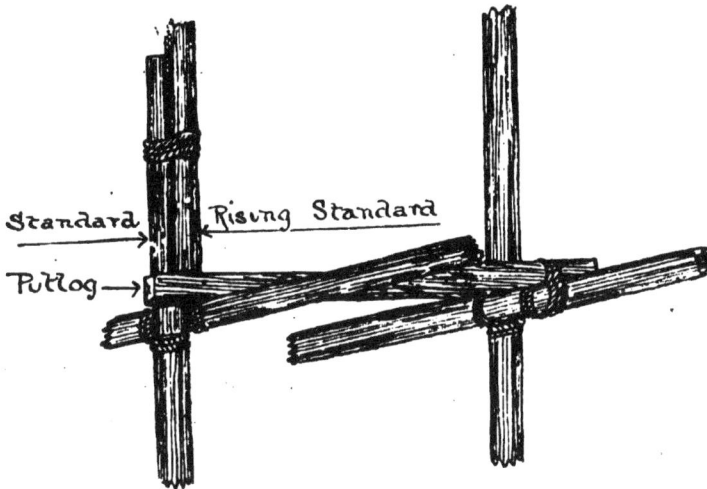

FIG. 14.—METHOD OF FIXING RISING STANDARD

As the standards rise, they are spliced or 'married'
together with band ties.

At a height of 5 feet, this distance being the greatest
at which a man can work with ease, a ledger is tied
across the standards to form a support for the working
platform.

Where a single pole is insufficient in length to form

a continuous ledger, two are joined in one of three ways.

In the first they are lapped over each other as fig.

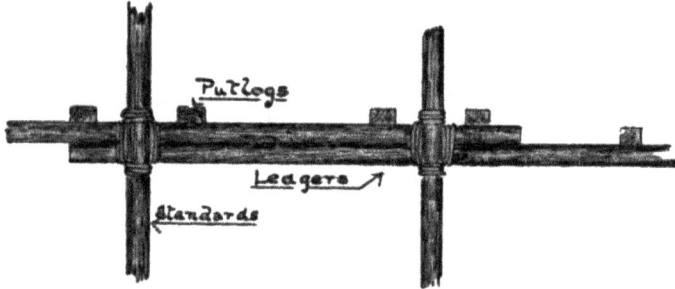

FIG. 15

15. This method gives a strong connection, but prevents the putlogs being laid evenly.

The second way provides that the ledgers shall lap horizontally side by side. This allows of evenness of line for the putlogs, but is not so strong (fig. 16).

FIG. 16

In both of these methods the lap should cover two standards, and not as shown in fig. 17.

The third manner of connection (fig. 18) is the best. The ledgers butt end to end. Underneath, a short pole is placed crossing two standards. The tying at the standard embraces the double ledger. A band tie is run round the supporting pole and the ends of the ledgers where they butt.

Great strength is obtained in this way and the putlogs can be evenly laid.

Additional ledgers are fixed as the work proceeds.

On the ledgers, and at right angles to them, putlogs

FIG. 17

are laid, resting outwardly on the ledgers and inwardly
on the wall, where header bricks have been left out for
their reception.

The putlogs, which are placed about 3 or 4 feet
apart, should be tied to the ledgers and fastened by

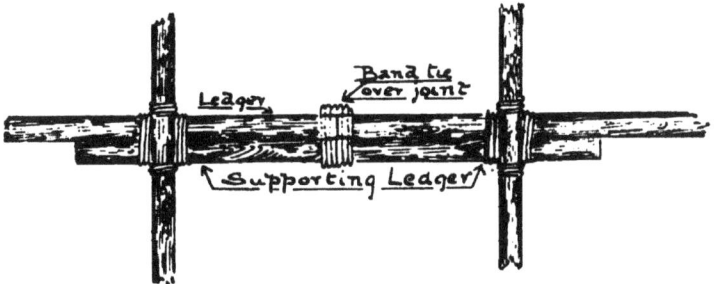

FIG. 18

wedges into the wall. This is not often done, but at
least one putlog to every tying between standard and
ledger should be so treated.

Where the putlogs cannot be carried by the wall

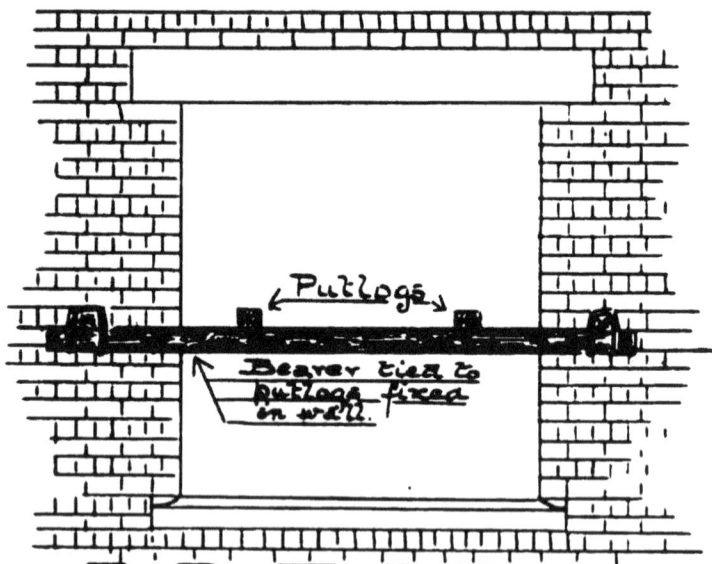

FIG. 19

FIG. 20

owing to an aperture in the building, such as a window, they are supported by bearers fixed as shown in figs. 19 and 20.

By wedging the inner end of the putlog into the wall, some stability is given to the scaffold, but the connection cannot be considered satisfactory, as the putlog would draw under. very little strain. Greater

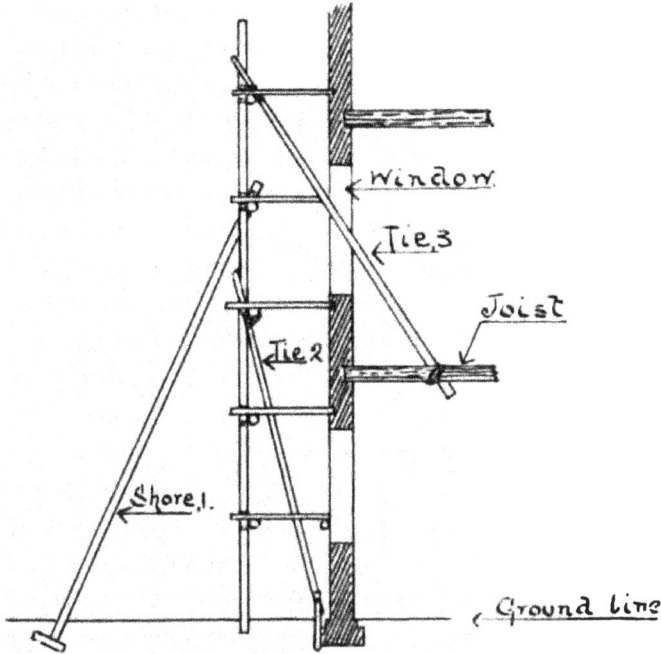

FIG. 21.—SHORES AND TIES FOR DEPENDENT SCAFFOLDS

stability can be gained if the outer frame of the scaffold is supported by one of the three methods given as follows.

A shore or tie can be fixed between the erection and the ground as shown in fig. 21, or, if there are openings in the wall, supports can be fixed as ties shown in the same diagram.

The ties or struts should be placed to every third or fourth standard at about 25 feet from the ground, and their fastenings made good. Additional ties should be carried within the building at a greater height where possible. The stability of the scaffold under loads and

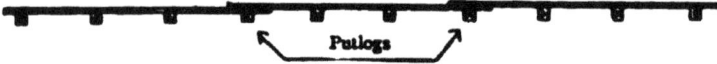

Putlogs

FIG. 22

cross wind pressure, depends greatly upon the ties or shores, and their fastenings should be well made and kept in good order. The historical instance of the mechanic who, to escape a shower of rain, stood upon the inner board of the platform, and by leaning against the building pushed the scaffold over, should have no opportunity of recurrence.

To stiffen the scaffold longitudinally braces are tied on the outside of the scaffold in the form of a St. Andrew's cross (see fig. 13).

They start from the lower end of one standard and rise obliquely across the scaffold to near the top, or some distance up a standard in the same run. Tied at their crossing-point, connections are made to all the main timbers of the scaffold with which they come in contact. Braces are fixed in all exposed situations, and

Putlogs

FIG. 23

generally where the scaffold is more than one pole (30 feet) in height.

The only exception to effective bracing being carried out is where the building, being of irregular form, creates many breaks and returns in the scaffolding. It is obvious

that where a scaffold butts against or breaks with a
return wall, the tendency to lateral motion is lessened.

The boards, which are placed longitudinally across
the putlogs, can be laid to lap or butt at their ends.
When lapping, one putlog only is required to carry the
ends of two series of
boards (fig. 22).

When butting, two
putlogs are required
placed about 4 inches
apart (fig. 23).

The second method
is the better, as the
boards are not so likely
to lose their place or to
trip the workmen. If
heavy work is in pro-
gress the boards are laid
double. As the build-
ing rises, the boards
are carried up to each
successive platform,
but each tied putlog is
left in its place.

Masons' Scaffolds.

—Masons', or indepen-
dent scaffolds differ
from the bricklayers' in
that they have to be

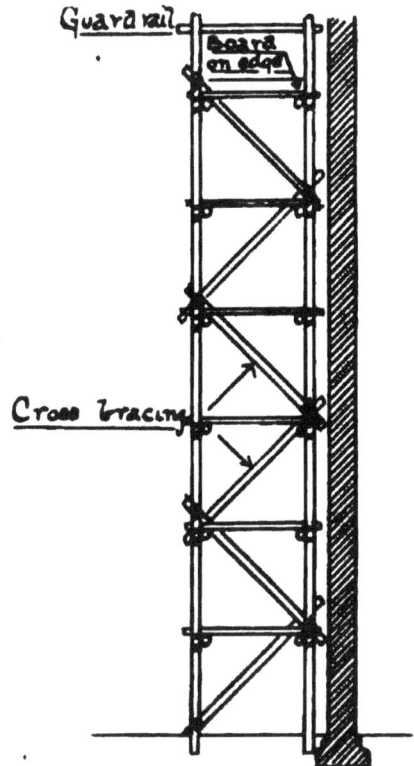

Fig. 24.—Masons' Scaffolds: End
Elevation

self-contained. Owing to the different material of which
the building is erected, the putlogs cannot rest upon the
wall. If openings were left for them, as in brick-work,
the wall would be permanently disfigured, more especially when ashlar fronted.

In order to gain the necessary support two parallel frames of standards and ledgers are erected along the line of wall to be built (fig. 24).

They are from 4 to 5 feet apart, the inner frame being as close to the wall as possible. As a heavier material has to be dealt with, the standards are placed closer together, say from 4 to 5 feet.

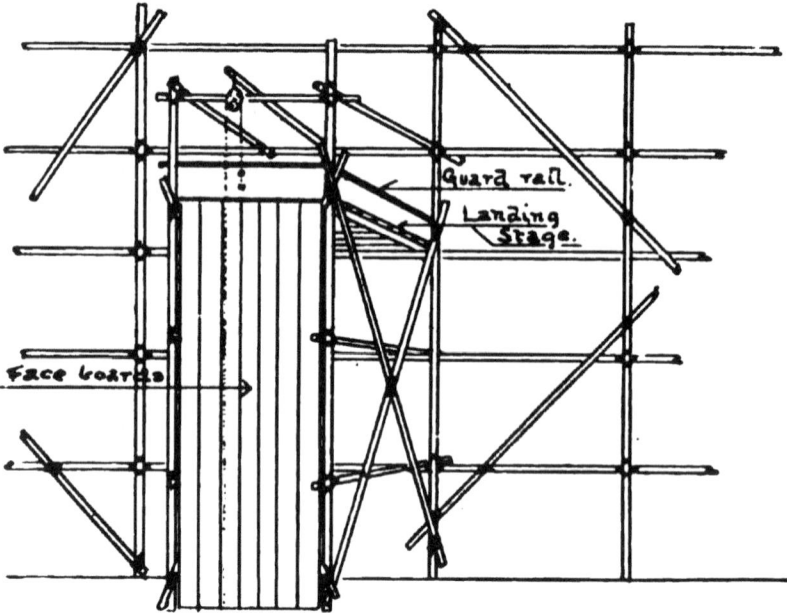

FIG. 25.—LANDING STAGES

The ledgers and braces are placed as before, the putlogs now resting on ledgers at each end, and not on the wall at the innermost end, as in the bricklayer's scaffold.

To prevent cross movement of the scaffold, an additional method of bracing is available in this system. An inner and outer standard are connected by short braces across each bay, as shown in fig. 24.

This method of cross-bracing can be continued to the top of the scaffold, and the braces should be put in longitudinally, about 20 feet apart.

The platforms laid on all pole-scaffolds are from 4 to 5 feet wide. It is usually necessary, on anything but the smallest jobs, to keep this width free for the workman and his material.

In order, therefore, to provide a platform on which the material can be landed, it is convenient to erect, on the outside of the scaffold, an additional platform from 5 to 10 feet square (fig. 25).

It is constructed of standards, ledgers, and braces, in like manner as the scaffold to which it is attached.

The face-boards, as shown in this figure, should be fixed wherever material is being hoisted, to prevent any projection of the load catching under a ledger and upsetting.

Connections.—The members of pole scaffolds are connected by cordage. The names of the various knots are given in Chapter V.

The arranging of the various timbers used in erecting scaffolds is a dangerous occupation, and one requiring skill and considerable nerve on the part of the workmen. In the majority of cases, the timbers on the ground level are placed in position by manual labour only, shear legs being used to facilitate matters. When the scaffold rises, advantage is taken of any rigid member on which pulley wheels can be hung, and by this means the succeeding poles, &c. are raised, manual dexterity and strength being responsible for their final position.

CHAPTER II .

SCAFFOLDS FOR SPECIAL PURPOSES

WHEN applying the given methods for scaffolding, difficulties arise owing to the varying designs of the buildings under construction or repair.

It is impossible to deal with these cases in detail ; they must be left to the scaffolder, who, while holding closely to the principles, by the exercise of ingenuity will make combinations and variations of the various systems to suit the special requirements demanded in each case. There are, however, certain types of scaffolding which occur with some regularity, and these will now be dealt with.

Needle Scaffolding.—Needle scaffolding is neces-sary where it is impossible or too expensive to carry the scaffold from the ground level or other solid base. It is used both for repairing and new erections.

The needles from which the scaffold takes its name are timbers (usually poles or balks) placed horizontally through and at right angles, or nearly so, to the wall of the building. The projections support a platform upon which an ordinary pole scaffold is erected (fig. 26).

Windows, or other openings in the wall, are utilised where possible for the poles to pass through. In other cases holes have to be made in the walls, cut as nearly as can be to the size of the needles in use.

The needles must be of sufficient scantling to carry

the weight of the scaffold and attendant loads. The stability of the structure depends upon the means taken to fasten down the inner end of the needle.

The usual plan is to tie it down to a convenient joist or other rigid member of the building itself, but the method shown on the diagram is better, as resistance to movement is gained both from above and below.

Struts from the building below the needles to their outer end, give greater strength to the beam.

When erecting needle scaffolding around buildings of small area, say of a tower or chimney shaft, the needles can be laid across the building in one length, piercing the wall on opposite sides. In these cases, if the needles are wedged in, the weight of the building and the scaffold itself on the opposite ends of the needles, is sufficient to maintain equilibrium.

FIG. 26.—NEEDLE SCAFFOLD

The platform is formed of 9-in. by 3-in. deals, and on this is erected whatever scaffolding may be necessary.

Scaffolds for Chimney Shafts, Towers, and Steeples.—The erection of chimney shafts can be

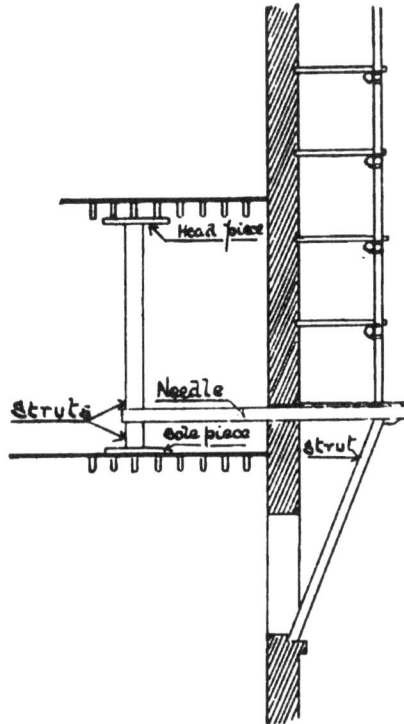

carried on entirely by the aid of internal scaffolding.
As the work rises putlogs are laid across the shaft,
the ends being well built into the wall. On the putlogs
the platform is laid, being carried up as the work
proceeds. The putlogs may be left in for the time, and
struck on completion. The platform is fitted in its
centre with a hinged flap door through which the
material is hoisted as required.

There is some objection to this method of scaffolding where the wall is more than 1 foot 10½ inches
thick (which is the greatest depth of brickwork over
which a man can reach and do finished work), for the
mechanics, in order to reach the outside joints, have to
kneel on the freshly laid material, which is detrimental
to good workmanship. For this reason the system of
carrying up an ordinary pole scaffold externally until
the height is reached where the wall is reduced to 1 foot
10½ inches in thickness, is to be preferred.

The walls of a chimney shaft decrease in thickness
4½ inches at a time, forming an internal set-back of that
width at every 20 feet in height.

This set-back is of advantage to internal scaffolding
when the full height of the brickwork is reached, and
the cap has to be fixed. The cap or coping, when of
stone or iron, does not admit of the insertion of putlogs.
To overcome the difficulty, four or more standards are
erected at equal distances, and standing upon the top
set-back (fig. 27).

The standards project sufficiently to carry the pulley
wheel well above the total height of the chimney, in
order to give head room and to assist the workman in
fixing the coping.

To stiffen the standards, short ledgers are tied across
as shown in fig. 27.

When the chimney is to be erected by external

EXTERNAL CHIMNEY SCAFFOLD.

Erected for the Willesden Electric Lighting Works, under the supervision of E. WILLIS, Esq., A.M.I.C.E., etc.

scaffolding the ordinary mason's or bricklayer's scaffold is used. Owing to the small area of the erection the outside frames of the scaffold have a quick return. This makes it practically impossible for the scaffold to fail by breaking away from the building under the influence of the loads it may carry. Shoring or tying is therefore not so important. Wind pressures have, however, a greater effect, especially when the direction is not at right angles to one of the faces of the scaffold. If in that direction, the tied putlogs would offer resistance. Braces are therefore imperative, and they should be fixed at right angles to each other, each pair thus bracing a portion of the height of the scaffold equal to its width. (See plate 2.)

For the repair of chimney shafts without scaffolding from the ground level, means have to be taken to bring, first the mechanic, and afterwards his material, within reach of the work.

The preliminary process of kite-flying is now rarely

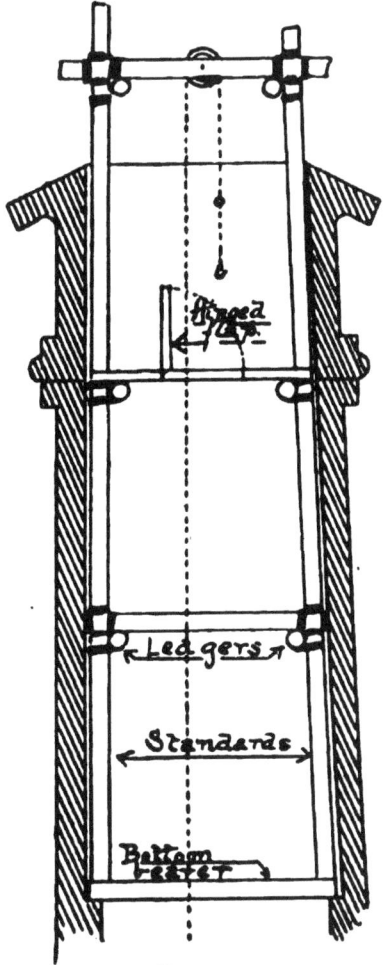

Fig. 27

seen, except for square-topped chimneys, and even in these cases the delay that may arise while waiting for a suitable steady wind is a drawback to its practice. The kites used are about 10 feet long and 8 feet wide. They are held at four points by cords which continue for a distance of about 16 feet, and then unite into one. Near this point on the single rope another cord is attached, which serves to manipulate the kite into position.

Stronger ropes or chains are then pulled over the shaft, after which a workman ascends, and the necessary pulley wheels and timbers to form a regular means of ascent are sent up after him.

A light line carried up in the interior of the shaft by a hot-air balloon is another means of communication.

The most certain and safest method of ascent is to raise on the exterior of the shaft a series of light ladders, which are lashed to each other and firmly fixed to the chimney as they ascend.

The ladders have parallel sides, and are used up to 22 feet in length.

One method of fixing is as follows :—

A ladder is placed against the shaft on its soundest side. It rests at its top end against a block of wood slightly longer than the width of the ladder, and which keeps it from 7 to 9 inches away from the wall. This space allows room for the workmen's feet when climbing. The ladder is then fixed by two hooks of round steel driven into the wall, one on each side immediately under the blocks, the hooks turning in and clipping the sides of the ladder (fig. 28). The hooks, which have straight shanks of $\frac{1}{2}$-inch diameter with wedge-shaped points, are driven well home, as the stability of the erection depends upon their holding firmly.

Above the top end of the ladder a steel hook is

driven into the wall on which a pulley block can be hung, or instead, a pin with a ring in its head can be so fixed. A rope from the ground level is passed through this block or ring, and reaches downward again for connection to the ladder next required. The connection is made by lashing the rope to the top rung and tying the end to the seventh or eighth rung from the bottom; this causes the ladder to rise perpendicularly. The steeplejack who is standing on the already fixed ladder cuts the top lashing as the hoisted ladder reaches him, and guides it into its place as it rises. When the rung to which the rope is tied reaches the pulley block, the ladders should overlap about 5 feet. They are at once lashed together at the sides, not round the rungs.

The workmen can now climb higher, driving in hooks round the sides, and under the rungs of the ladder alternately, lashings being made at each point. A wooden block is placed

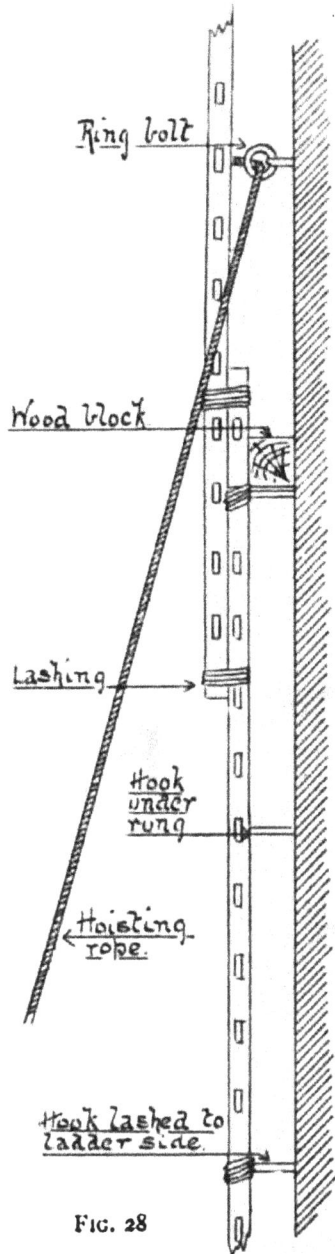

Ring bolt

Wood block

Lashing

Hook under rung

Hoisting rope.

Hook lashed to ladder side.

Fig. 28

under the top end of the last ladder and fixed as before. The hoisting rope, which has been kept taut meanwhile, is now loosened and the process repeated.

The ladders rise in this manner until the coping of the shaft is reached. Here, owing to the projection of the cap which throws the ladders out of line, it is impossible to lash the top ladder to the lower. To overcome the difficulty, the wall is drilled in two places immediately over the topmost fixed ladder, and expansion bolts are fitted therein. To these bolts the lower end of the top ladder is tied. The hoisting rope is then tightened sufficiently to hold the ladder, and by this means the workmen are enabled to reach the top of the shaft.

A variation of this method of climbing is to replace the wooden blocks by iron dogs with 9-inch spikes, which should be driven well into the wall. Short ladders of about 10 feet in length are then used, these being lashed to the dogs as they rise.

Another method of fixing the ladders is shown in fig. 29.

In this case eye-bolts are driven horizontally into the wall in pairs, rather wider apart than the width of the ladders.

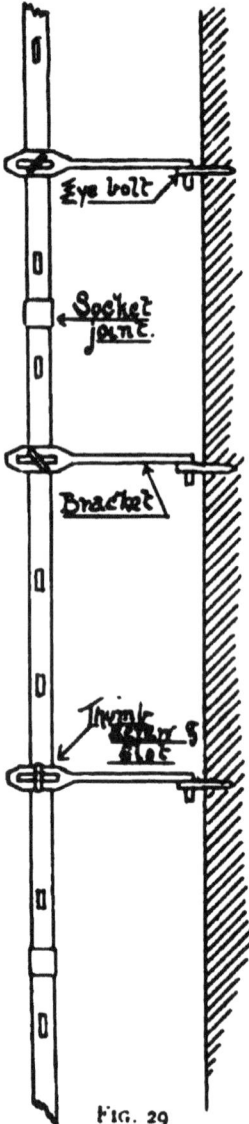

Eye bolt

Socket joint.

Bracket

Thumb screw slot

FIG. 29

Iron rods hook into these and are fastened to the ladder sides by thumb screws.

The ladders rise above each other and are connected by 3-inch sockets.

When fixed, they stand about 18 inches from the wall.   This is an advantage, as it enables the workmen to climb on the inside of the ladders, thus lessening the strain on the eye-bolts, and the ladder can more easily pass a projecting chimney cap.

On the other hand, the whole weight of the ladders rests upon the bottom length, so that if through any cause it gave way, for instance under accidental concussion, the entire length would most certainly collapse.

This danger could be avoided if the ladders were supported on brackets as fig. 30. No reliance should be placed upon the thumb screws, as they may work loose under vibration. Danger from this source would

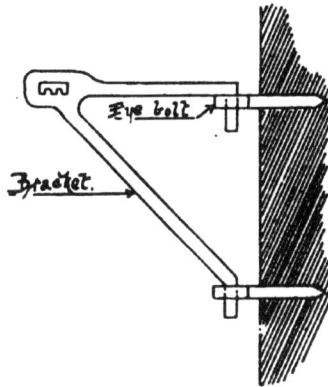

Fig. 30

be avoided if the slot in which the ladder peg moved was made as shown in fig. 30.

The necessary repairs can be carried out by means of boats, cradles, or scaffolding.

Cradles and boats are swung from balk timbers laid across the top of the shaft, or from hooks where the design of the chimney permits, as shown in fig. 31.

The common method of fixing light scaffolds round a chimney or steeple is shown in fig. 32.   They are most easily fixed to square or other flat-sided erections. The scaffolder having by means of ladders or boats reached the desired height, fixes a putlog by means of holdfasts to one of the walls. Another putlog is then fixed on the opposite side of the building at the same level. The two are next bolted together by 1-inch iron bolts of the required length. The bolts are kept as near to the wall as possible. The process is repeated again about 6 feet higher on the building. The boards for the platforms are next laid. The first are placed at right angles to the putlogs and project sufficiently to carry the boards which are laid parallel to the putlogs. To prevent the boards rising when weight is applied at one side of the scaffold, iron plates bolted together (fig. 33) are fixed at the corners, and clips (fig. 34) connect them to the putlogs.

The stability of these scaffolds depends upon fixing

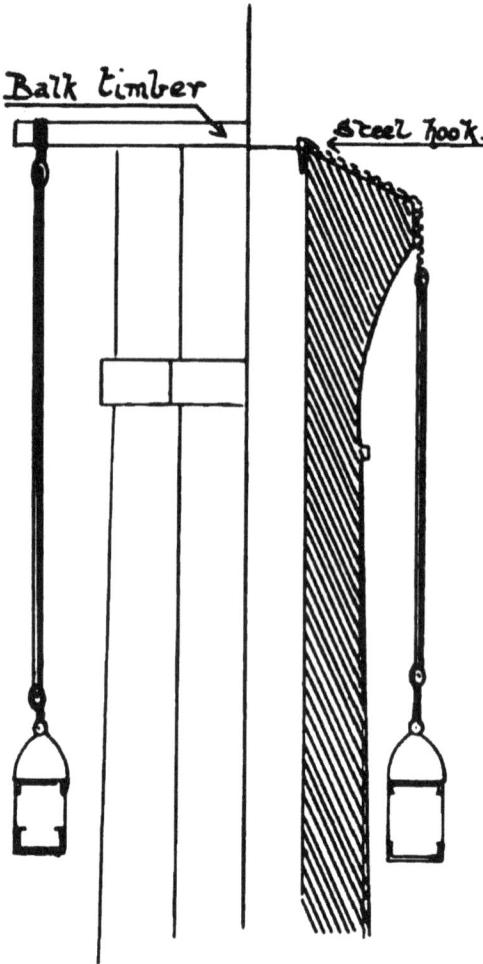

Balk timber

Steel hook.

FIG. 31

at least two sets of putlogs, connected by means of stays as shown in fig. 32.   Bracing is unnecessary if the put-logs and bolts tightly grip the building.   When these

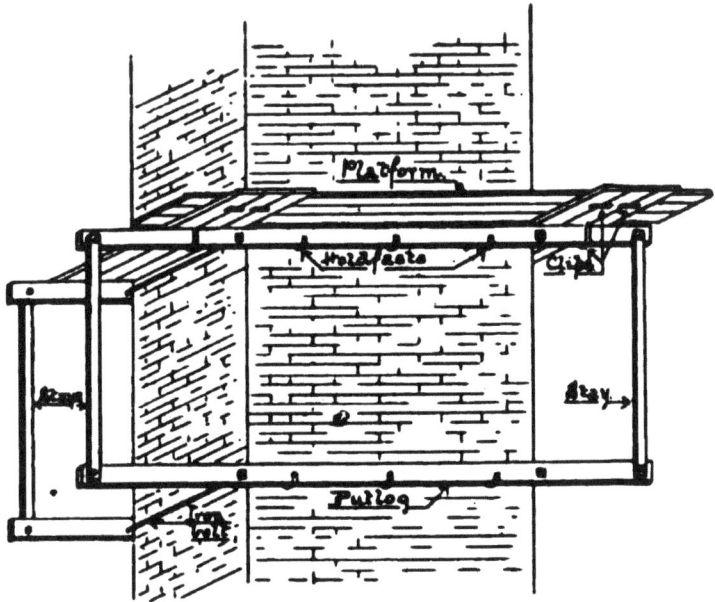

FIG. 32

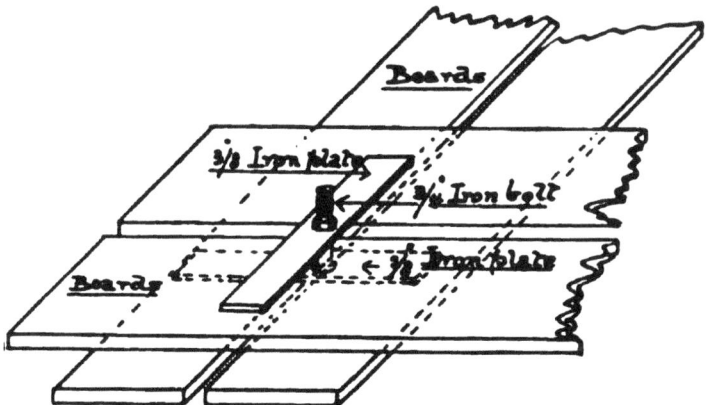

FIG. 33

scaffolds are used on circular chimneys, chucks have to be fitted on the inside of the putlogs to prevent them being drawn by the bolts to a curve. The chucks (fig. 35) can be fastened to the putlogs before they are fixed, if the curve of the building is accurately known. When this is not the case, the putlogs are fixed by a holdfast at their centre. The

FIG. 34

chucks are then placed in position, and clamped to the putlogs as shown in fig. 36.

Additional holdfasts are then driven into the wall immediately under the chucks, so that the putlogs are kept level.

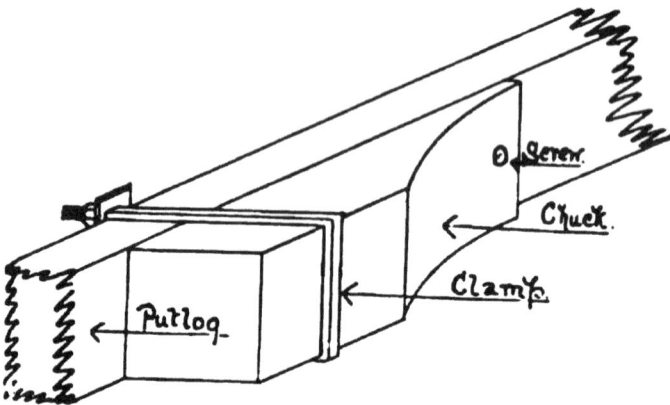

FIG. 35

The putlogs are fixed on edge, and when not ex- ceeding 16 feet in length are 7 in. by 3 in. Above that length they are 9 in by 3 in. The stays should be 4 in. by 2 in., and connected to the putlogs by

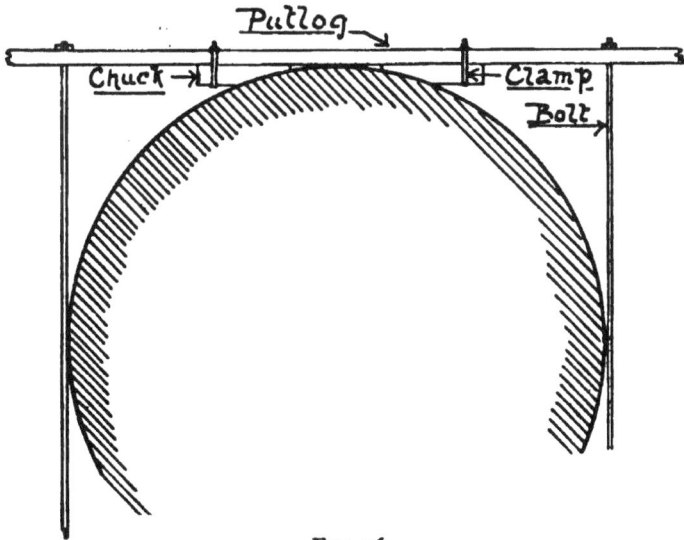

Fig. 36

⅝-inch iron bolts. The platform is usually of three boards 11 in. by 2 in.

Hollow towers are erected or repaired in the same manner as chimney shafts, except that climbing ladders are not often required. External or internal scaffolds may be erected. Towers being usually of larger area than chimney shafts, the putlogs for internal scaffolding are often of short poles from 6 to

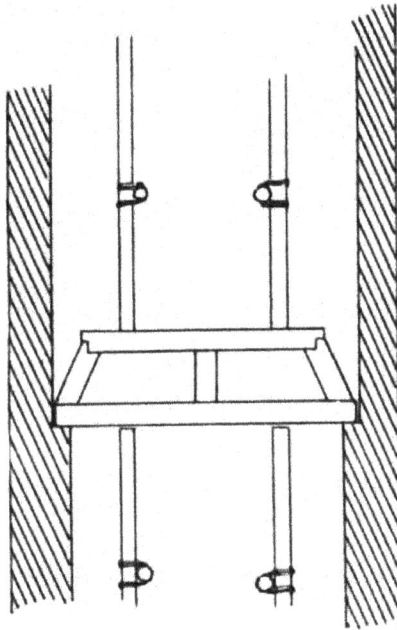

Fig. 37

8 inches diameter. Even these may require extra support. This is gained by carrying standards from the
ground level or other solid foundation and tying to the
putlogs. If of great height the standards may be unable
to carry their own weight. For the cases where danger
might be apprehended from this cause, fig. 37 shows a
system of framing, which, being supported by the set-
back in the thickness of the wall, will carry the upper
standards.

Steeples are generally built by the aid of external
scaffolds, which, as in the case of chimney shafts,
should be well braced. The lower portion may also be
repaired in this way, the standards rising from the ground
level, or, if so designed, from the top of the tower. A series
of needles could be arranged for the higher portions.

Domes and arches.—The scaffolding for domes and
arches consists of a series of standards standing upon

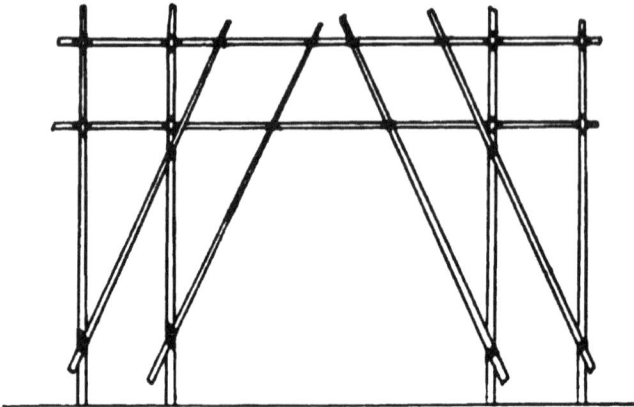

Fig. 38

the area covered by the building, and connected by
ledgers and braces in directions at right angles to each
other. The platform is laid on the top ledgers.

When the building is of large span square timbers are often used, balks for standards and runners, and half timbers for struts and braces.

Fig. 38 shows a design for repairing roofs and arches where a roadway has to be kept below.

Swinging scaffolds. Painters' boats or cradles.— Painters' boats are useful scaffolds for the repair of buildings, more especially where the work is light. Fig. 39 shows the general construction. They are

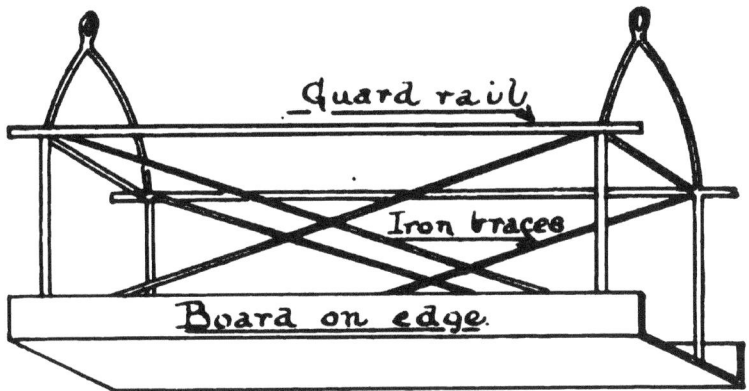

FIG. 39

suspended from jibs, fixed usually on the roof for out-side work, and by means of blocks and falls they can be moved in a vertical direction by the workmen when in the boat.

The boats are fitted with guard boards and rails, and their safety, providing the jibs are well fixed by balancing weights, is in their favour. They are not self-supporting, and there is a distinct danger of their running down if the sustaining ropes are not securely fastened off. The wind causes them to sway consider-ably, and their use is confined chiefly to façade work.

An improved cradle is now in general use, which is slung by head blocks from a wire cable running between

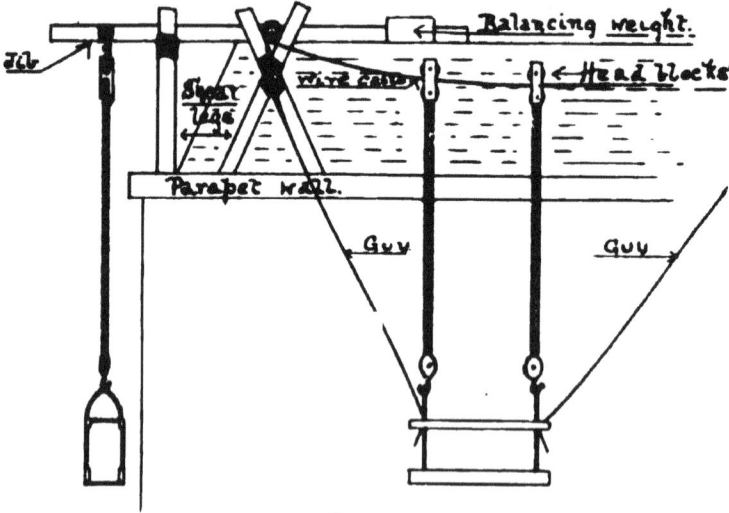

FIG. 40

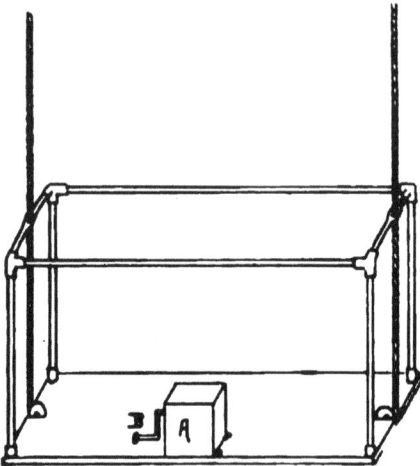

FIG. 41

two jibs (see fig. 40). By the aid of guy lines movement in this case can be also obtained horizontally, which removes the necessity of shifting the jibs or employing a greater number of boats as in the older method.

Another cradle as shown in fig. 41 has advantages which cannot be ignored. It has steel cables with a breaking weight of 15 cwt. instead of fibre ropes, and the cradle is raised and lowered by means of gearing and a drum fixed in the gear case A. It is self-supporting, and therefore safer than the cradle mentioned above. The lower ends of the cable are fastened to the drum, and the gearing gives sufficient mechanical advantage for one man to raise the scaffold by turning the handle B. The uprights and rails are of angle steel or barrel and will take apart and fold.

FIG. 42

The boatswain's boat (see fig. 42) is useful under some circumstances, especially for making examinations of buildings for possible damage. It is dangerous and awkward to work from, and is also acted upon considerably by the wind.

The boat is slung from a single needle. The workman has no control over its movement, as he has to be raised or lowered as required by men having charge of the other end of the fall.

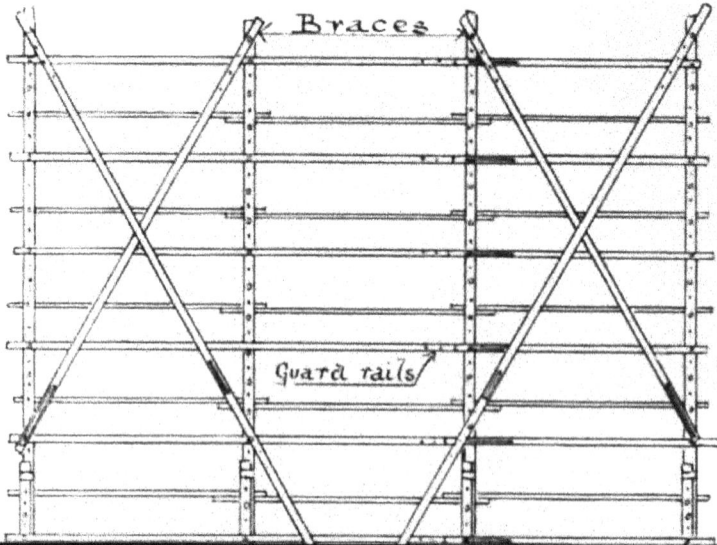

purpose of guard rails, is shown in fig. 43.   The ladders, which have parallel sides, are placed about 2 feet

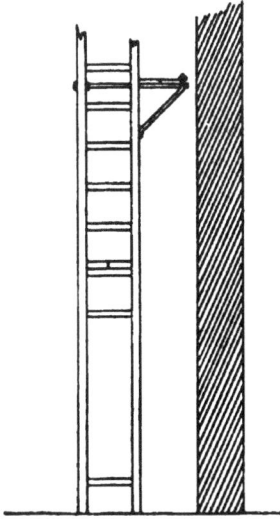

FIG. 44

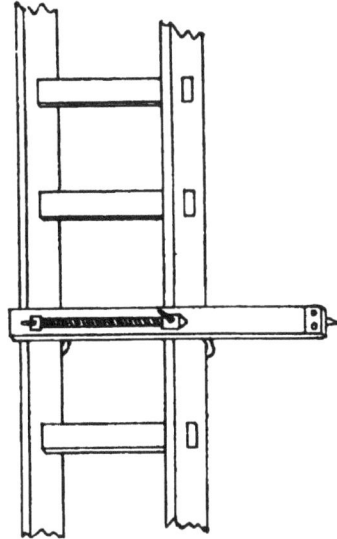

FIG. 45

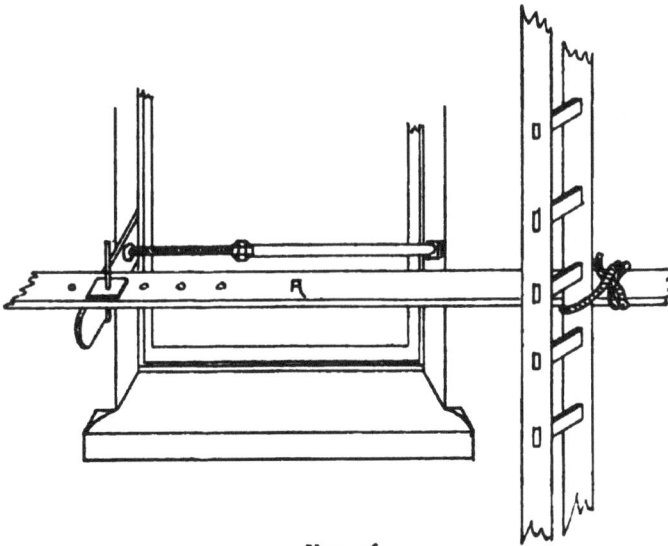

FIG. 46

away from the building. The boards forming the platform can be laid on the ladder rungs, or if necessary on brackets as shown in fig. 44. The ladders are prevented from falling away from the building by ties which are connected to the ladder as shown in fig. 45, and fastened to the window openings by extension rods as shown in fig. 46. The same figure illustrates the method of tying in the scaffold when the ladders are not opposite to the windows, the rail ʌ being connected to at least two ladders. The braces and guard rails are bored for thumb screws at one end, the other being slotted so that they can be adjusted as required. This form of scaffold is only suitable for repairing purposes, and no weight of material can be stored upon it.

A light repairing scaffold lately put on the market has a platform which is supported and not suspended, but otherwise affords about the same scope to the workmen as the painters' boats. It consists of one pole and a platform, the latter being levered up and down the pole as required by a man standing on the platform itself. The whole apparatus can be moved by one man standing at the bottom. It is an arrangement comparatively new to the English trade, but is in considerable use in Denmark, Germany, and Sweden.

CHAPTER III

SHORING AND UNDERPINNING

SHORING is the term given to a method of temporarily supporting buildings by a framing of timber acting against the walls of the structure. If the frame consists of more than one shore, it is called a system ; if of two or more systems, it becomes a series.

The forces that tend to render a building unstable are due primarily to gravity, but owing to the various resistances set up by the tying together of the building, the force does not always exert itself vertically downwards.

This instability may arise from various causes, the most common being the unequal settlement of materials in new buildings, the pulling down of adjoining buildings, structural alterations and defects, and alterations or disturbances of the adjacent ground which affect the foundations. The pulling down of an adjoining building would, by removing the corresponding resistance, allow the weight of the internal structure of the building to set up forces which at first would act in a horizontal direction outwards. Structural defects, such as an insufficiently tied roof truss, would have the same effect. Structural alterations, such as the removal of the lower portion of a wall in order to insert a shop front would allow a force due to gravity to act vertically downwards.

E

To resist these forces, three different methods of shoring are in general use, and they are known as flying or horizontal shores, raking shores, and underpinning.

Flying Shores.—Where the thrusts acting upon the wall are in a horizontal direction, flying or raking shores are used to give temporary support. The most direct resistance is gained by the first-named, the flying or horizontal shore. There are, however, limits to its application, as, owing to the difficulty of obtaining sound timber of more than 50 to 60 feet in length, a solid body is necessary within that distance, from which the required purchase can be obtained.

It is a method of shoring generally used where one house in a row is to be taken down, the timbers being erected as demolition proceeds, and taken down again as the new work takes its place.

Fig. 47 shows a half-elevation of two general systems of construction.

The framing, as at A, may be used alone where the wall to be supported is of moderate height and the opening narrow, but larger frames should be combined, as at B.

The framework C is for wide openings and walls of considerable height.

The wall plates, 9 in. by 2 in. or 9 in. by 3 in., are first fixed vertically on the walls by wall hooks. Then, in a line with the floors, rectangular holes 4 in. by 3 in. are cut in the centre of wall plates. Into these holes, and at least 4½ inches into the brickwork, needles (also known as tossles and joggles) of the same size are fitted, leaving a projection out from the wall plate of 5 in. or 6 in., sufficient to carry the shore of about 7 in. by 7 in.

The shore, prior to being fixed, has nailed on its top and under sides straining pieces 2 inches thick, and of the same width as the shore. To tighten, oak folding wedges are driven at one end between the shore and wall plate.

To stiffen the shore, and to further equalise the given resistance over the defective wall, raking struts are fixed

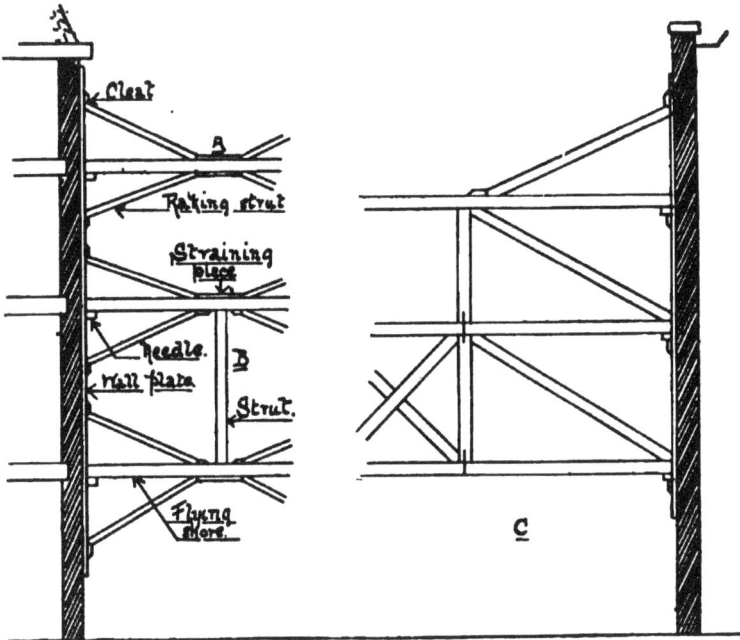

FIG. 47

between the straining pieces, and cleats are nailed above and below the shore. These raking struts are tightened by driving wedges between their ends and the straining pieces.

The cleats previous to, and in addition to being nailed, should be slightly mortised into the wall plate. This lessens the likelihood of the nails drawing under the pressure.

A **Raking Shore** consists of a triangulated system of timber framing, and is used to support defective walls where the resistance to the threatened rupture has to be derived from the ground surrounding the building.

In its simplest form a raking shore is a balk of timber of varying scantlings, but as a rule of square section, inclined from the ground to the defective wall. The angle of inclination is taken from the horizon, and should vary between 60 and 75 degrees. In settling this the space available at the foot of the wall has to be taken into consideration, especially in urban districts where the wall abuts on the footpath.

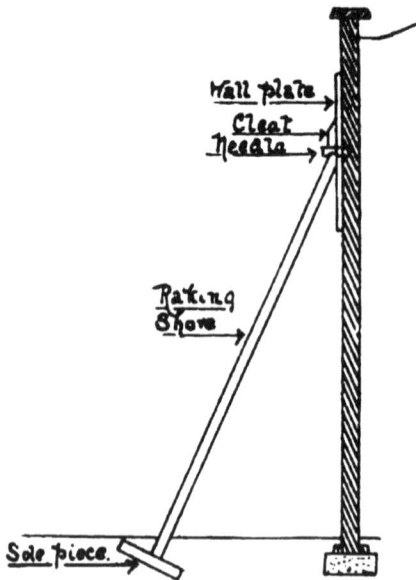

Wall plate
Cleat
Needle
Raking Shore
Sole piece.

Fig. 48

Fig. 48 shows a raking shore in its simplest form, but usually two or more shores are used (see fig. 49).

The following table from Mr. Stock's book [1] shows the general rule and also the scantlings to be used :

For walls from 15 to 30 feet high 2 shores are necessary in each system

30 „	40	„	3	„	„
40 and upwards			4	„	

[1] *A Treatise on Shoring and Underpinning, and generally dealing with Dangerous Structures.* By Cecil Haden Stock. Third edition, revised by F. R. Farrow. (B. T. Batsford.)

Rider or
Top shore

Top middle
raker

Strut

Bottom
raker

Wall
plate

Folding wedges

93°

Bottom
shore

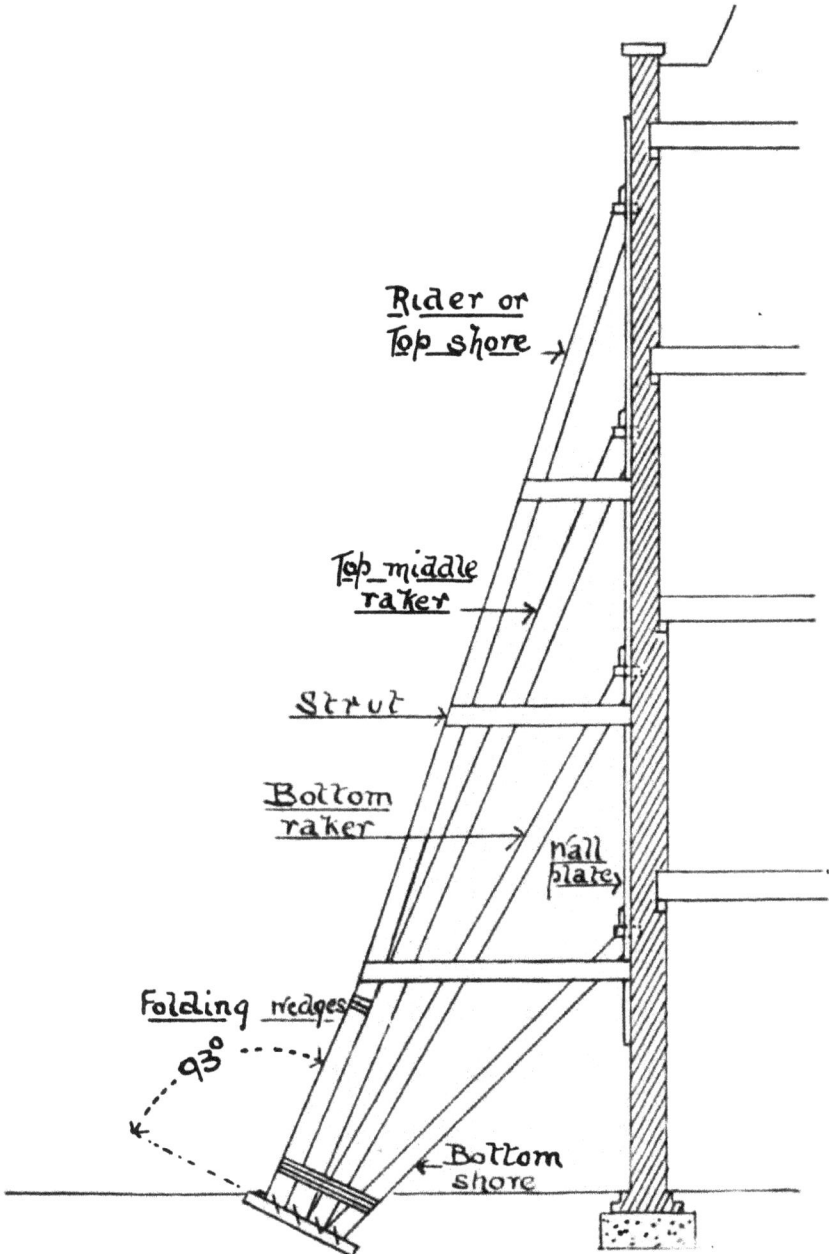

FIG. 49.—EXAMPLE OF RAKING SHORE

Taking the angle of the shore at from 60 to 75 degrees :

For walls from 15 to 20 feet high 5 in. by 5 in. may be the scantling
for each shore

,,	20	,,	30	,,	6	,,	6	,,	,,
,,	30	,,	35	,,	7	,,	7	,,	
,,	35	,,	40	,,	8	,,	8	,,	
,,	40	,,	50	,,	9	,,	9	,,	
,,	50 and upwards 12	,,				,,	9	,,	,,

In the greatest length, the beams are 12 in. by 9 in. to give increased rigidity, which prevents any likelihood of sagging.

The wall plate is the first timber put into position. It is placed vertically down the face of the wall, and held in its position by wall hooks. Note should then be taken of the position of floors. If the floor joists run at right angles to the wall, the shore should abut on the wall in such position that it points directly below the wall plate carrying the floor joist. If the joists run parallel to the wall, the shore should act directly on a point representing the meeting of lines drawn down the centre of the wall and across the centre of the floor (see fig. 50).

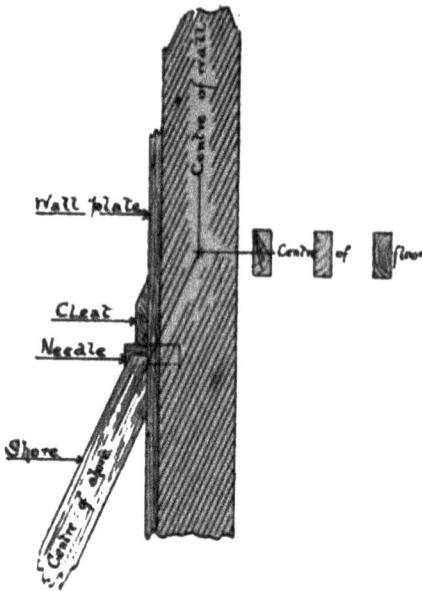

FIG. 50

To enable the shore to fulfil this condition, the needle (of 4 in. by 3 in.) should be let through the plate 4½ inches into the wall below the point in question. To strengthen the needle cleats are nailed, and slightly let into the plate immediately above.

The footing, or sole piece, has next to be laid. It consists of a timber 11 in. by 3 in., and long enough to take the bottom ends of the required number of shores. Attention should be paid to the ground in which it is to be bedded, and if this is at all soft, additional timbers should be laid under, and at right angles with it, to give greater bearing.

The sole piece should not be laid at right angles to the shore, but its face should form, with the outside line of the top shore, an angle somewhat wider, say of 93 degrees. The advantage of this will be seen presently.

The shore itself has now to be prepared. Its top end should be grooved sufficiently (fig. 51) to receive the needle. This will prevent lateral motion when under pressure.

FIG. 51

The bottom end should be slightly slotted, in order to receive the end of a crowbar (see fig. 52).

It is now placed in position, and gently tightened up by the leverage of a crowbar acting in the slot, and using the sole piece as a fulcrum.

The advantage of the sole piece not being at right angles to the shore can now be seen, as if it were so laid no tightening could be gained by the leverage. This system is an improvement upon the tightening up by wedges, as the structure is not jarred in any manner. If the frame is to have more than one shore, they are

erected in the same manner, the bottom shore being the
first put up, the others succeeding in their turn. When
in position the shores are dogged to the sole piece and
a cleat is nailed down on the outer side of the system.
The bottom ends are then bound together by hoop iron
just above the ground level. To prevent the shores
sagging, struts are fixed as shown on fig. 49.

Besides preventing the sagging these struts serve
the purpose of keeping the shores in position. They

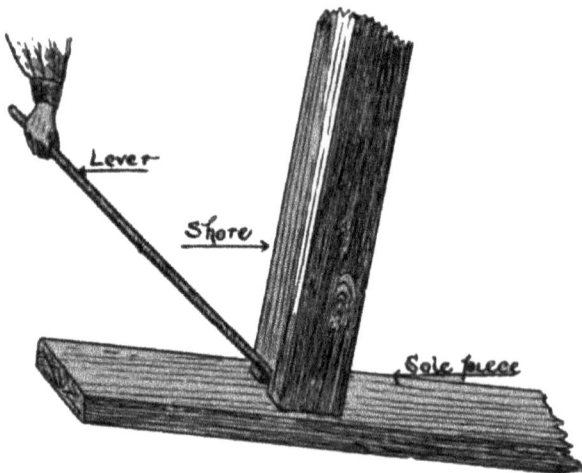

FIG. 52

may be fixed as nearly at right angles to the shores as
possible, or at right angles to the wall ; in any case they
should reach to the wall plate at a point just below the
needle. The struts should be nailed to the shores and
wall plate. If the latter is wider than the shores, it
should be cut to receive the struts.

It sometimes occurs that the timbers are of insufficient length to reach from the sole piece to wall
plate. To overcome this difficulty, a short timber is
laid on the sole piece against and parallel to the next

middle raker, and on this short timber a rider shore stands reaching to its position on the wall plate (see fig. 49).

When this is done the top middle raker should be stiffer to resist the increased cross strain. Stiffness is gained by increasing the depth. A rider shore is tightened by oak folding wedges driven between the foot of the shore and the short timber which supports it.

Note must be taken that the outer raker is not carried too near the top of the building, or else the upward thrust of the shores, which always exists with raking shores, might force the bond or joints.

Fir is the best wood for shoring owing to the ease with which it can be obtained in good length. Another advantage is its straightness of fibre; although, as it is more easily crushed by pressure across the grain, it does not answer so well as oak for wedges, sole pieces, &c.

In erecting flying or raking shores, notice should be taken of the following points.

The systems should be placed from 12 to 15 feet apart if on a wall without openings, otherwise on the piers between the openings.

In very defective walls it is an advantage to use lighter scantlings, the systems being placed closer together. Heavy timbers handled carelessly may precipitate the collapse which it is the intention to avoid.

Wedge driving and tightening should be done as gently as possible. It should be remembered that support only is to be given, and not new thrusts set up, which may result in more harm than good.

Underpinning.—Underpinning is necessary to carry the upper part of a wall, while the lower part is removed ; for instance, the insertion of a shop front, or the repairing of a foundation. It is only kept in position until a permanent resistance to the load is effected. Underpinning is, as a rule, unnecessary when the opening to be made is of less width than five feet. This method of shoring is a simple operation, but yet requires great care in its execution.

The first thing to be done is to remove from the wall all its attendant loads. This is accomplished by strutting from the foundation floor upwards from floor to floor until the roof is reached (see fig. 53).

Header and sole plates 9 in. by 2 in. are put in at right angles to the joists in order to give bearing to the struts.

The portion of the wall to be taken down having been marked out, small openings are made, slightly above the proposed removal, at from 5 to 7 feet apart, and through these, at right angles to the face of the wall itself, steel joists or balk timbers 13 in. by 13 in., called needles, are placed. These are supported at each end by vertical timbers 13 in. by 13 in., called dead shores, which again rest upon sleepers.

The sleepers serve as a bed to the dead shores to which they are dogged, and by distributing the weight over a larger area, they prevent the dead shores sinking under the pressure. The dead shores, if well braced, may be of smaller scantling.

Where it is impossible to arrange for the dead shores to be in one length, the lower pieces are first fixed. They must be of uniform length, and across their top end a transom is carried to support the upper pieces, the bottom ends of which must stand directly over the top ends of the lower pieces (see fig. 53).

Having placed all the timbers in position, and before the tightening up takes place, the windows or

other openings in the wall are strutted to prevent any
twisting which may take place. This is done as shown

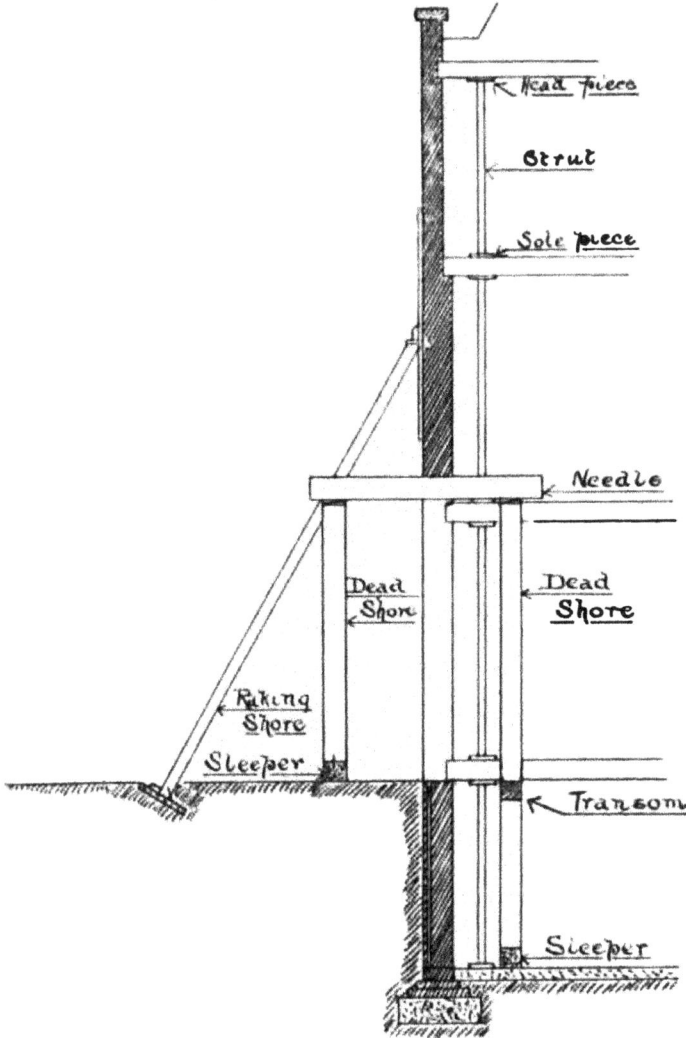

FIG. 53.—EXAMPLE OF UNDERPINNING

on fig. 54, but small windows do not require the
centering.

The tightening up is caused by the driving home of oak folding wedges placed in the joints between the needles and the dead shores. This position is better than between the shore and sleeper, as any inequality of driving here would have the tendency to throw the shore out of the perpendicular. For a similar reason the wedges should be driven in the same line as the run of the needle, as cross driving, if unequal, would cause the needle to present an inclined surface to the wall to be carried.

FIG. 54

In carrying out these operations note should be taken of the following points :—

1. That the dead shores should not stand over cellars or such places. It is better to continue the needle to such a length that solid ground is reached, and the needle can then be strutted from the dead shore.

2. That extra needles should be placed under chimney breasts, should there be any in the wall which is to be supported. The same applies where corbels, piers, &c. occur.

3. Sleepers and shores should be so placed that they do not interfere with the proper construction of the new foundations or portions of the building.

4. The inside shores should be carried freely through the floors until a solid foundation is reached.

5. The removal of the shores after the alterations have been made is one requiring great care. It should be remembered, that while the work is new it cannot offer its greatest resistance to its intended load. Time should therefore be given for the work to well set, and then the timbers, eased gradually by the wedges being loosened, should be finally taken out.

6. The raking shores, if used in conjunction with underpinning, should be left to the last.

CHAPTER IV

TIMBER

Classification and Structure.—A short study of the classification and structure of wood will be useful, as it will enable the scaffolder to use material free from the inherent defects of its growth.

The trees used by the scaffolder are known as the exogens or outward-growing trees.

The cross section of an exogenous tree shows, upon examination, that the wood can be divided into several parts. Near the centre will be seen the pith or medulla, from which radiate what are known as the medullary rays. These in the pine woods are often found full of resinous matter.

Next will be noted the annual rings forming concentric circles round the pith. They are so called because in temperate climates a new ring is added every year by the rising and falling of the sap. As the tree ages, the first-laid rings harden and become what is known as duramen or heartwood. The later rings are known as alburnum or sapwood. The distinction between the two is in most trees easily recognised, the sapwood being lighter and softer than the heartwood, which is the stronger and more lasting. The bark forms the outer covering of the tree.

Defects in the living tree.—Shakes or splits in the interior of the wood are the most common defects in the living tree, and are known as *star* or *radial* and *cup* or *ring shakes*. The cause of these defects is imperfectly understood. They are rarely found in small trees, say those of under 10 inches diameter. Stevenson, in his book on wood, puts forward the following reason, which, up to now, has not been refuted by any practical writer. ' In the spring, when the sap rises, the sapwood expands under its influence and describes a larger circle than in winter. The heartwood, being dead to this influence, resists, and the two eventually part company, a cup or ring shake being the result' (see fig. 55, where the cup shake is shown in its commonest form).

FIG. 55   FIG. 56

The star or radial shake is a variant of the same defect. In this case the cohesion between the sapwood and the heartwood is greater than the expansive forces can overcome, the result being that the heartwood breaks up into sections as shown in fig. 56.

The star shake may have two or more arms. More than one cup shake, and sometimes both cup and star shakes are found in a single tree. The radial shake is probably the most common.

The branches suffer in this respect in like manner as the trunk, the same shakes being noticeable throughout

The development of these defects is the forerunner of further decay in the tree, giving, as they do, special facilities for the introduction of various fungi, more especially that form of disease known as the rot. Wet rot is found in the living tree and occurs where the timber has become saturated by rain.

Other authorities believe that these defects are caused by severe frost, and their idea receives support from the fact that in timber from warmer climates this fault is less often seen. This may be so for the reason that it would not pay traders to ship inferior timber from a great distance.

Loose hearts : in the less resinous woods, as those from the White Sea, the pith or medulla gradually dries and detaches itself, becoming what is known as a loose heart. In the more resinous woods, such as the Baltic and Memel, this defect is rare.

Rind galls are caused by the imperfect lopping of the branches, and show as curved swellings under the bark.

Upsets are the result of a well-defined injury to the fibres caused by crushing during the growth. This defect is most noticeable in hard woods.

It will be of no practical use to follow the living tree further in its steady progress towards decay, but take it when in its prime, and study the processes which fit it for the purpose of scaffolding.

Felling.—The best time to fell timber, according to Tredgold, is mid-winter, as the vegetative powers of the tree are then at rest, the result being that the sapwood is harder and more durable ; the fermentable matters which tend to decay having been used up in the yearly vegetation. Evelyn, in his ' Silva,' states that : ' To make excellent boards and planks it is the advice of some, that you should bark your trees in a fit season, and so let them stand naked a full year before felling.' It is questionable if this is true of all trees, but it is often done in the case of the oak. The consensus of opinion is that trees should be felled in the winter, during the months of December, January and February, or if in summer, during July. Winter felling is probably the better, as the timber, drying more slowly, seasons better.

The spruce from Norway and the northern fir are generally cut when between 70 and 100 years old. When required for poles, spruce is cut earlier, it having the advantage of being equally durable at all ages. Ash, larch and elm are cut when between 50 and 100 years old.

Conversion.—By this is meant cutting the log to form balks, planks, deals, &c. It is generally carried out before shipment. A log is the trunk (sometimes called the stem or bole) of a tree with the branches cut off.

A balk is a log squared. Masts are the straight trunks of trees with a circumference of more than 24 inches. When of less circumference they are called poles.

According to size, timbers are classed as follow :—

Balks	12 in. by 12 inches to 18 in. by 18 inches
Whole timber . .	9 „ „ 9 „ „ 15 „ „ 15 „
Half timber . .	9 „ „ 4½ „ „ 18 „ „ 9 „
Quartering . .	2 „ „ 2 „ „ 6 „ „ 6 „
Planks . . .	11 „ to 18 „ by 3 „ to 6 „
Deals . . .	9 „ „ 2 „ „ 4½ „
Battens . . .	4½ „ „ 7 „ „ ⅞ „ „ 3 „

When of equal sides they are termed die square. In conversion the pith should be avoided, as it is liable to dry rot.

When the logs are to be converted to whole timbers for use in that size, consideration has to be given as to whether the stiffest or the strongest balk is required. The stiffest beam is that which gives most resistance to deflection or bending. The strongest beam is that which resists the greatest breaking strain. The determination of either can be made by graphic methods.

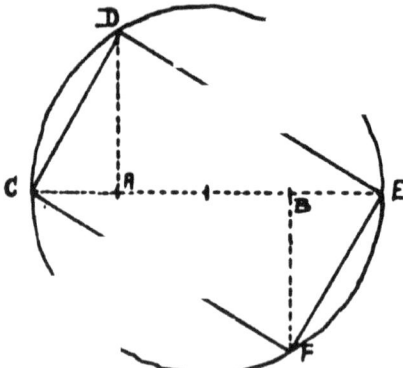

FIG. 57

To cut the stiffest rectangular beam out of a log, divide the diameter of the cross section into four parts (see fig. 57). From each outside point, A and B, at right angles to and on different sides of the diameter, draw a line to the outer edge of the log. The four points thus gained on the circumferential edge, C, D, E, F, if joined together, will give the stiffest possible rectangular beam that can be gained from the log.

To cut the strongest beam :—In this case the diameter is divided into three parts. From the two points A and B thus marked again carry the lines to the outer edge as before. join the four points C, D, E, F, together, and the outline of the strongest possible rectangular beam will result (see fig. 58).

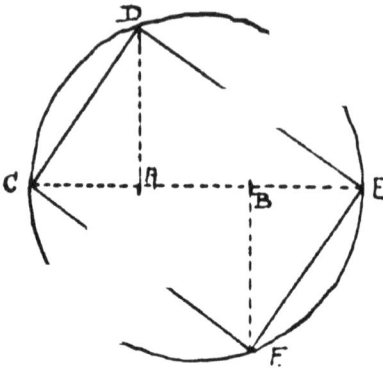

FIG. 58

Boards—a term which embraces planks, deals, and battens—should be cut out of the log in such a manner that the annual rings run parallel to the width of the board. This method of conversion allows the knots, which are a source of weakness, to pass directly through the board, as A, fig. 59, and not run transversely across as other sections (B) allow.

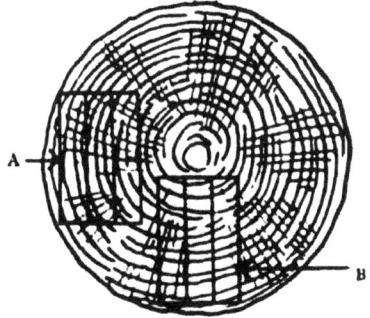

FIG. 59

Seasoning.—Poles and scantlings, after conversion, are next prepared for seasoning.

The drying yard, where the timbers are seasoned, should give protection from the sun and high winds, although a free current of air should be allowed. It should be remembered that rapid drying tends to warping and twisting of the timbers.

The ground on which the timbers are stacked should be drained and kept dry.

The method of stacking is as follows : The timbers are laid upon supports a few inches above the ground at sufficient intervals to allow of a free circulation of air between them, and on these others are laid at right angles. It is usual to put long strips of wood about ¾ inch square between each row to prevent the timbers touching, as, after a shower, the timbers cannot dry for a long time if one is resting on the other.

Planks, deals, &c. can be stacked by laying them in one direction throughout, provided that the space which is left between the boards occurs in one layer immediately over the centre of the boards beneath (see fig. 60).

In many cases very little drying takes place before shipment, but the same methods of stacking should be observed whenever it takes place.

Stacking poles by placing them on end is not recommended, as they may warp from insufficient

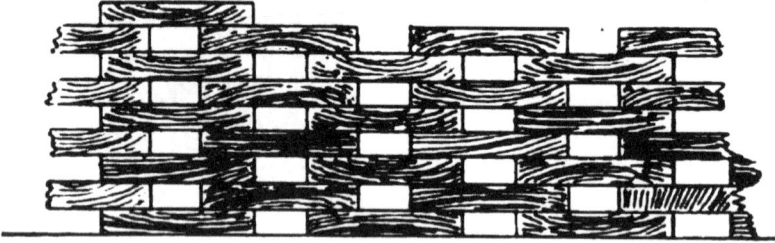

FIG. 60

support. This point is more important when the poles are required for ladder sides.

Timber may be considered to be sufficiently seasoned for rough work when it has lost one-fifth of its weight by evaporation.

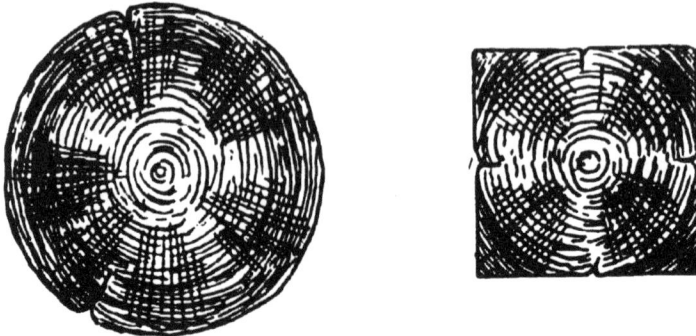

FIG. 61

The poles from St. Petersburg have a narrow strip of bark removed in four equidistant longitudinal lines throughout their entire length. This treatment assists drying, and tends to prevent dry rot.

Weather shakes sometimes form on the outside of the wood while seasoning (see fig. 61).

They arise owing to the sapwood contracting more when drying than the heartwood. Unless they extend to a considerable depth they do not affect the quality of the wood. Balk timbers, where the sapwood is uncut, and whole timbers principally suffer in this manner.

Water seasoning—that is, having the timbers completely immersed in water for a short time before drying. This is a common practice. It is frequently carried out at the docks, where the balks may be seen floating about on the surface of the water. This is a bad method, as the wood is at the same time under the influence of water, sun and air.

Water seasoning may make the wood more suitable for some purposes, but Duhamel, while admitting its merits, says: 'Where strength is required it ought not to be put in water.'

Description.—Pine or northern fir (*Pinus sylvestris*) is light and stiff, and is good for poles and scaffolding purposes, but only the commonest of the Swedish growths are used for this purpose. The hardest comes from the coldest districts. It has large red knots fairly regularly placed, inclined to be soft, and starting at acute angles. It grows chiefly in Northern Europe.

White spruce or white fir (*Abies excelsa*).—The knots are small and irregularly placed, are dark in colour, and start at an obtuse angle, showing an absence of clean wood. It is used principally for scaffold poles and ladder sides. It will snap under live loads, and is not so strong as pine. It also chiefly grows in Northern Europe.

Larch (genus *Larix*) is imported from Northern Europe and America. It is yellow in colour, tough, and is suitable for poles, very durable, free from knots, but warps easily.

Elm (genus *Ulmus*) grows in the British Isles. The colour varies from the reddish-brown of the heartwood to yellowish white of the sapwood. It bears considerable pressure across the grain, and is most useful in balks, as it is liable to warp when in smaller sizes. Is suitable where bolts and nails have to be used.

Birch (*Betula alba*).—Light brown in colour, hard, even grain, which enables it to be easily and readily split in the direction of its length. It is not considered durable. It is used chiefly for putlogs. Is exported from Europe and America.

Ash (*Fraxinus excelsior*).—Very light brown in colour, extremely tough, and makes excellent rungs for ladders. It is found in Europe.

Oak (genus *Quercus*), also known as the common British oak, is a native of all parts of Europe, from Sweden to the Mediterranean. The wood has often a reddish tinge ; and the grain is fairly straight, and splits easily. It is generally free from knots, and is most suitable where a stiff, straight-grained wood is desirable. It also offers considerable resistance to pressure across the grain.

Selection.—The importation of scaffolding timber commences in March. In June the Russian timbers are on the market, and the arrivals continue until October.

Poles are selected from spruce, pine, and larch trees. Balk timbers are of elm and fir and spruce. The putlogs are from the birch.

Poles are known in two qualities—'prime' and 'brack.' These terms refer to their straightness of grain, freedom from knots, regularity of taper, which should be slight, and condition as to seasoning.

Buyers take them usually unbarked, as they rise from the stack, and sort them afterwards for their different purposes.

As soon as possible the bark should be removed, as it holds the water and insects, and encourages the growth of a white fungus which is the precursor of dry rot.

The St. Petersburg poles are the most generally used for scaffolding. They are weaker than the Christiania poles, breaking shorter under cross strains, are whiter in colour, have a smoother bark, are straighter in grain, and therefore make better ladder sides.

Christiania poles are more yellow in colour, and break longer under a cross strain than other poles. They are only to be obtained early in the year, and are soon bought up when on the market.

To test a pole remove the bark, then prise across the grain with a penknife. If the fibres break up short and brown, the pole is decayed and useless.

To test a pole for any local weakness lift one end, leaving the other on the ground. Two or three sharp jerks will cause undue bending at any spot that may be seriously defective.

Balk timbers ring, if in a sound condition, when struck with a hammer. A fresh-cut surface should be firm, shining, and somewhat translucent. A dull, chalky, or woolly appearance is a sign of bad timber.

Poles, unlike most converted timbers, are not branded. Those known as *prime* should certainly be used for scaffolding purposes. The brack are rough and irregular in growth.

Balk timbers and smaller scantlings are, in the finest qualities, branded, the different countries from which they are exported being Russia, Norway and Sweden,

Russian woods are generally hammer branded, no colour being used.

Norwegian woods are marked a blue colour, and Swedish wood a red colour.

Straight-grained timber should be chosen. Twisting, which may occur to the extent of 45°, is not so apparent in young trees. The beam is thereby weakened as regards tension and compression.

Large knots should be avoided. They weaken the beam for tension and cross strains, but serve good purpose when in compression.

Sap in the wood is denoted by a blue stain, and betokens inferiority as to strength and lasting qualities. This blueness is more noticeable when the wood is wet. It must not be confounded with dark weather stains.

The Decay of Timber.—Timber exposed to the constant changes of weather tends to decay early. It has been noticed that wood when dried after being exposed to dampness not only lost the moisture it had absorbed, but also part of its substance. This loss occurs in a greater degree when these changes take place a second time. It is therefore apparent that scaffolding timber, exposed as it is to the vicissitudes of the weather, without any protection such as would be gained by painting, will soon from this cause show signs of decay.

Quicklime, when wet, has also a most destructive effect upon timber ; the lime, by abstracting carbon, helping the decomposition materially. Scaffold boards suffer most in this respect from contact with fresh mortar, the lime in which has not had time to become mild.

From a similar cause, the ends of putlogs which are

inserted in a newly built wall, as when used in a brick-layer's scaffold, tend to decay rapidly.

Scaffolding poles, when used as standards, have an increased tendency to decay at their butt ends, owing to their being imbedded in the ground to a distance of two or three feet.

The presence of sap in improperly seasoned wood is also conducive to early decay.

Preservation of Timber.—Scaffolding timber being comparatively cheap has little attention paid to it in regard to preservation.

When out of use the poles should be stacked, and a free current of air around each ensured, not laid care-lessly on the ground, which is too frequently the case. The same remarks apply to the care of boards. Nails which have been driven into the timber should, after use, be carefully drawn. When left in the wood they rust, and set up a new source of decay.

As before noted, the butt ends of standards tend to decay early. A fairly effective method of preventing this is to bore the butt end upwards to a distance of about 2 feet. The bore, which should be about $\frac{1}{2}$ inch in diameter, should then be filled with pitch and plugged. The pitch will find its way by capillarity into the pores of the wood and tend to keep the dampness out. Coating the outsides of the butts with pitch is also useful, and a combination of the two would no doubt be effectual ; this is especially recommended where the standards have to remain in one place for a long time.

Ladders and fittings that are expensive are gene-rally painted.

The Durability of Wood.—In Young's ' Annals of Agriculture ' it is stated that experiments were made on some 1½-inch planks of 30 to 45 years' growth. They were placed in the weather for ten years, and then examined, with the following result :

Larch—heart sound, sapwood decayed.

Spruce fir—sound.

Scotch fir—much decayed.

Birch—quite rotten.

This experiment, while useful to show the natural resistance of the wood to the weather, does not take into account the effect of wear and tear.

It is almost impossible, in fact, to arrive at any data from which the life of scaffolding timbers can be gauged ; but, roughly speaking, poles may be expected to last from six to ten years according to the care exercised. Balk timbers, being usually cut up after a time for other purposes, have only a short life on a scaffold, and therefore seldom decay while in use.

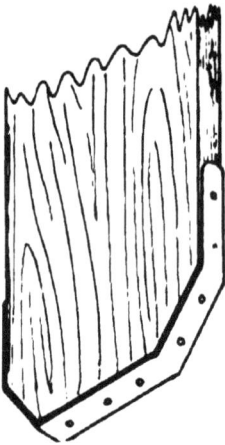

The Use of Scaffolding Timber.—Poles vary up to 40 and 50 feet in length and up to 8 inches in diameter at the butt end. As decay, which usually commences at this end, sets in, the poles can be shortened and made into sound puncheons or splicing pieces for the ledgers. Balks are used up to 50 or 60 feet in length ; and their period of service on a scaffold is often an interval during which they become well seasoned and suitable for other requirements. Putlogs are about 6 feet in length and 4 inches square in cross section, tapering sharply to 2¼ by 3¼ inches at the ends where required for insertion in the wall. Being of square section they are not liable

FIG. 62

to roll on the ledgers. They should be split, not sawn, in the direction of their length. The fibres are thus uncut and absorb moisture less easily. This procedure it is found increases their durability. When treated in this manner they are also stronger, as the fibres, being continuous, give greater resistance to a load, the strength not depending wholly on the lateral adhesion, but also upon the longitudinal cohesion. When partly decayed they can be shortened and make good struts for timbering excavations.

The boards are usually from 7 in. to 9 in. wide and 1¼ in. to 2 in. thick. Their length averages up to 14 feet. The ends are sawn as fig. 62 and strapped with iron to prevent splitting. When decayed they are a source of danger for which there is only one remedy — ·smash them up.

CHAPTER V

CORDAGE AND KNOTS

THE fibres used for the ropes and cords for scaffolding purposes are of jute and hemp. The strongest rope is made of the latter variety, and the best quality is termed the Manilla.

The cords are generally tarred for preservation, but this process has a bad effect upon the strength of the fibre, more especially if the tar is impure. The process of tarring is as follows :—

The fibres before they are formed into strands are passed through a bath of hot tar. Immediately afterwards, and while still hot, the material is squeezed through nippers, by which means any surplus tar is removed.

The ropes, as used in scaffolding, are known as shroud laid or three strand.

A shroud-laid rope consists of three strands wound round a core, each containing a sufficient number of fibres to make an equal thickness. A three-strand rope is similar, but has no core.

The following gives the breaking weights of scaffold cords and falls according to circumference [1] :—

	Size		Cwts.	qrs.	lbs.
Tarred hemp scaffold Cords	2 inches in circumference		20	0	0
	2¼ "	"	25	0	0
Tarred jute scaffold Cords	2 "		14	0	0
	2¼ "		18	0	0
White Manilla scaffold Cord	2 "		32	0	0
	2¼ "		39	0	0
	2¾ "		44	0	0
White Manilla rope for pulleys and block ropes	3 "		70	0	0
	3½ "		98	0	0
	4 "	"	126	0	0
	4½ "	"	158	0	0

[1] As derived from tests made by Messrs. Frost Bros., Ltd., rope manufacturers, London, E.

It should be noted that all these weights are the actual breaking strains of the ropes tested to destruction. In practice, one sixth of the breaking weights only should be allowed, in order to leave a sufficient margin for safety.

The strength of a rope is the combined strength of each separate yarn. Therefore, if the fibres are not carefully twisted together, so that each bears an equal strain, the rope is unsafe. Care should be taken when choosing a rope that the strands are closely, evenly and smoothly laid.

The rope is strongest when the fibres are at an angle of 45 degrees to the run of the rope. When at a greater angle than this, the fibres are apt to break, and when at a less angle, the friction between the parts—upon which the strength of a rope greatly depends—is lessened.

The durability of tarred ropes is greater than when untarred, but not to such an extent if kept dry.

A splice weakens a rope about one eighth.

Ropes are adulterated by the admixture of rubbish fibre termed 'batch.' To test a rope as to condition, untwist it and notice whether any short ends break upwards. If so, or if the tarring has decayed internally, the rope must be viewed with suspicion.

Hemp ropes, after four to six months' wear, are often one fourth weaker than new ones.

Scaffold cords are from 15 to 18 feet in length.

Moisture will cause a shrinkage of 6 inches in an 18-foot cord.

A good hemp rope is more reliable than an iron chain, as the latter sometimes snaps on surgeing.

The following diagrams show the various knots used in scaffolding, both in the erection of the scaffold and for attaching materials to a hoist. They are shown in a loose condition as being more useful for study by a pupil. Several knots are, however, too intricate to explain by diagram, and in these cases an attempt has been made, with but scant success, to instruct as to their method of creation in the notes. It will, however, be found by the student that half an hour with an expert scaffolder, preferably an old sailor, will afford more instruction than hours spent in studying diagrams.

The art of knot making is governed by three principles. First, a knot must be made quickly. Secondly, it must not jam, as this prevents it being undone easily. Lastly, under strain, it should break before slipping.

Plate III.—No. 1. *Overhand or thumb knot.* Prevents the end of a rope opening out or passing through the sheaves of a block.

No. 2. *Figure of eight knot.* Used as No. 1.

No. 3. *Square or reef knot.* Will sometimes jam with small ropes.

No. 4. The knot we all make until we learn better, known as the granny, and will both slip and jam.

No. 5. *The bend or weaver's knot,* used for joining ropes together or securing a rope through an eye splice.

No. 6. *Wolding stick hitch,* is serviceable only in connection with a pole used as a lever.

No. 7. *Bale sling,* for hanging on to hook of lifting tackle.

No. 8. *Magnus hitch, or rolling hitch,* for lifting material.

No. 9. *Two half hitches, or builder's knot, or clove hitch.* Used for tying ledgers to standards.

No. 10. *Loop knot,* used where ends of rope are not available.

No. 11. *Loop knot,* used where ends of pole are not available.

PLATE III.   79

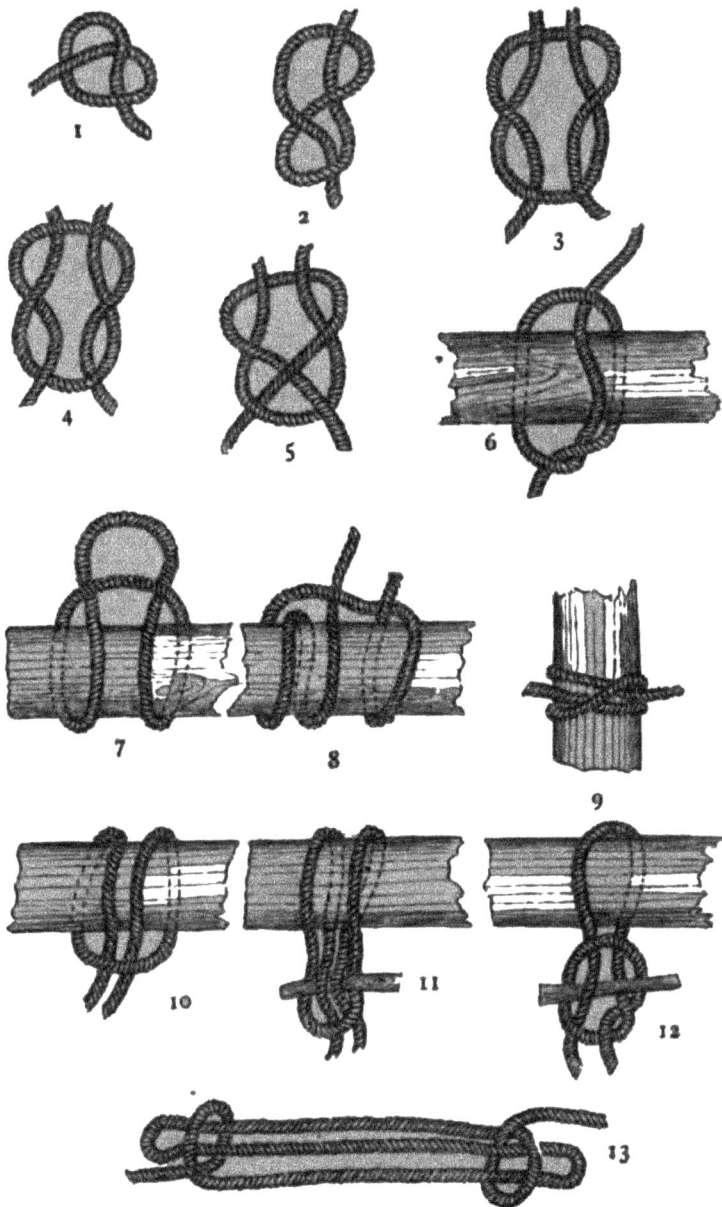

PLATE IV. 81

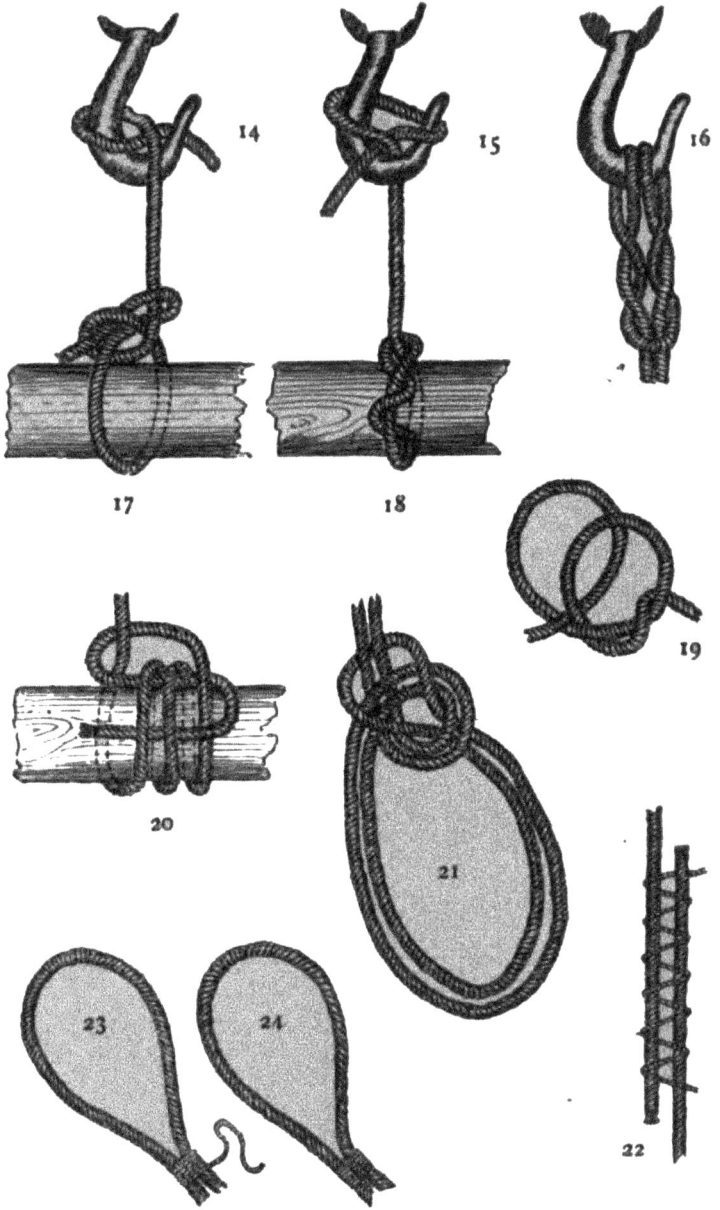

14

15

16

17

18

19

20

21

22

23

24

No. 12. *Boat knot.*

No. 13. *Sheepshank, or dogshank*, a method of shortening a rope without cutting it or reducing its strength.

Plate IV.—No. 14. *Blackwall hitch*, very powerful, but requires watching or may slip.

No. 15. *Midshipman's hitch*, used as shown with a rounded hook.

.No. 16. *Catspaw* is an endless loop, and is used where great power is required.

No. 17. *Capstan knot, or bowline.* After tightening it will not slip.

No. 18. *Timber hitch*, for carrying scaffold poles. Take one turn round the pole and standing part, and finish with jamming turns.

No. 19. *Artificer's knot*, or half hitch and over-hand.

No. 20. *Topsail halliard bend*, used as a timber hitch.

No. 21. *Bowline on a bight.* A board across the loops makes a useful seat.

No. 22. *Racking or nippering* is a method of temporarily joining two ropes for lengthening purposes. The ends are laid side by side for about 18 inches, and the marline or spunyarn is taken for about a dozen turns round both, then by round turns over all and fastened with a reef knot.

Nos. 23 and 24. *Round seising.* With a slip knot at the end of the spunyarn take a turn round the ropes to be nipped together. This turn should be pulled tight, and continued for about a dozen turns (No. 22); then take the end through the last turn, and take turns over the first laid, finishing by carrying the spunyarn two or three times between the rope and the seizing. Knot the

end by jamming turns, keeping the whole well taut
(No. 24).

Plate V.—No. 25. *Butt or barrel sling* when
placed horizontally.

No. 26. *Butt or barrel sling* when placed vertically.

No. 27. *Double overhand knot.*

No. 28. *Running bowline.*

Nos. 29 and 30. *Band tie, marrying or splicing.*
Commence as No. 29, and after continuing the turns
until near the end of the rope, take the rope twice
between the poles and round the turns first laid, and
finish with jamming turns. Tighten with a wedge.

Nos. 31, 32 and 33. *Tying between standard and ledger*
Commence with two half hitches as No. 31. Then
twist ropes together as far as they will go, and place
ledger in position above the hitches (No. 32). The
twisted ropes are then drawn up in the front of the
ledger to the left of the standard, taken round the back
of the standard, brought again to the front and round
ledger at the right of the standard, then cross in front of
the standard and round the ledger at the left of the
standard, and brought up and carried round the back of
the standard. This process is repeated until the end of
the rope is nearly reached, when it is given two or three
turns round the ledger and fastened off with jamming
turns (No. 33). To tighten, drive a wedge at back of
standard.

Plate VI.—No. 34. *Portuguese knot* for shear legs,
made by several turns of the rope round the poles, and
interlaced at the ends.

No. 35. *Running bow knot*—inferior to No. 28.

No. 36. *Bowline*—inferior to No. 17.

No. 37. *Double bend*—useful where a small rope

PLATE V.

85

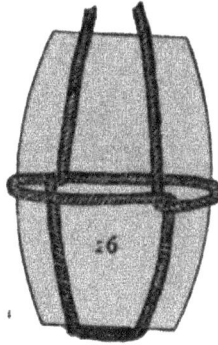

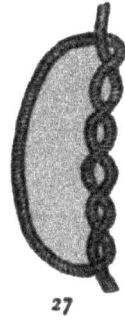

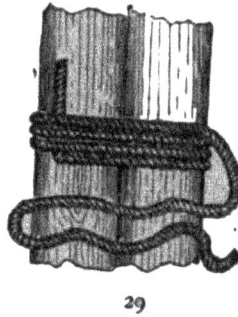

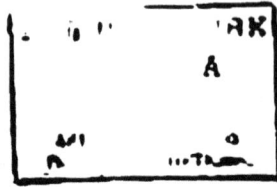

PLATE VI.   87

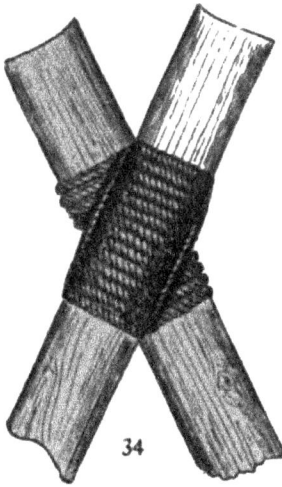

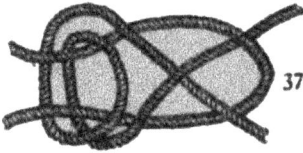

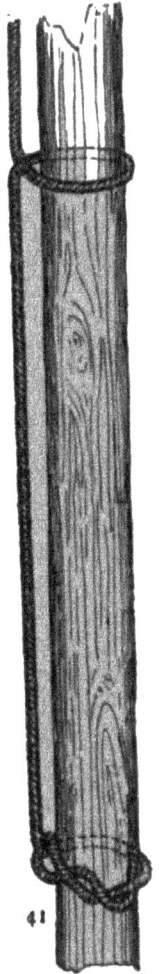

is bent on to a larger. The end of the rope is given one extra turn round the bight of the other, with the consequence of a great increase of strength.

No. 38. *Fisherman's knot.*

No. 39. *Lark's head,* fastened to a running knot.

No. 40. Where increased strength is required a small rope can be attached to a larger one by means of a rolling hitch. The whole arrangement comes apart as soon as the strain is removed.

No. 41. A method of lifting scaffold poles in a vertical position by the use of the timber hitch and half hitch. If it is required to free the upper end while the pole is being carried, the half hitch can be replaced by a cord tied round the pole and the lifting rope.

sides greatly weakens them ; but as the wedges may work out, an iron rod $\frac{5}{16}$-inch in diameter should be placed below every eighth or ninth rung, and bolted on the outside for extra security.

The iron rods should not be used as treads instead of the wooden rungs as they offer an insecure foothold. The rungs are considered to be dangerous for use when they have been re-duced by wear to one half of their original depth. The best rungs are made from old wheel-spokes, as they are well seasoned. The sides, which may be of sufficient length to receive 100 rungs, are 9 inches apart at the top

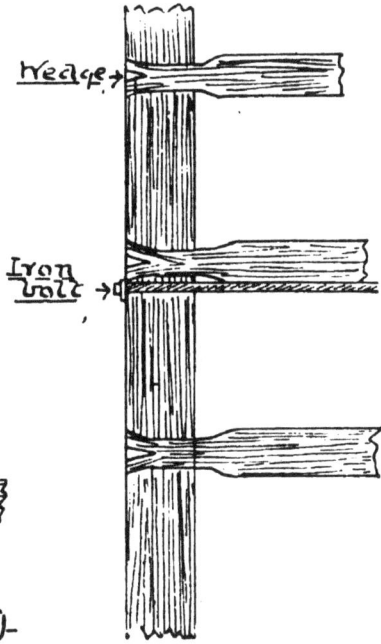

FIG. 64

FIG. 63

and from 12 to 13 inches apart at the bottom, according to the length of the ladder.

Extension ladders are useful where, owing to the varying heights of the work, different lengths of ladders are required. The two halves of the ladder are connected in various ways, but if well made they are easily raised and lowered. They should be used only for the very lightest work, such as painting, cleaning down, &c.

Trestles.—Trestles are used chiefly by painters plasterers, and mechanics engaged on work that is not at a great height from the ground or floor level, and for which a platform is required. They stand from 6 to 12 feet in height, and the rungs should be sufficiently wide to carry three boards for the working platform. They are made of yellow deal, with mortised joints and wrought-iron hinges (fig. 65).

FIG. 65

Steps.—Steps are built up with two sides of the required height, about 5 inches wide and 1 inch thick ; the top and bottom are sawn to a bevel so that they stand inclined.

The steps, which are grooved into the sides and fixed with screws, are about 6 inches wide by 1¼ inches thick, and increase slightly in length as they descend. This increase adds to the stability of the steps as the width of the base is increased. The distance between each step is from 7 to 9 inches.

At the back of the top step two legs about 2¼ inches wide by 1 inch thick are secured by strong flap hinges. The legs are framed together by two cross pieces, 3 or 4 inches wide and 1 inch thick.

The back legs, by opening out on the flap hinges, enable the entire framework to stand upon an even surface. To prevent the legs opening too far, they are connected to the sides of the steps by cords.

Cripples.—The simplest form of cripple is shown on fig. 66, which sufficiently explains the design.

This cripple forms a fixed angle with the ladder, which, in order to keep the platform level, can be laid only at one slope against the wall. The defect is removed if the cripple is hinged and fitted with a quadrant and pin, as shown on fig. 67. The platform in this case can be kept level by adjustment irrespective of

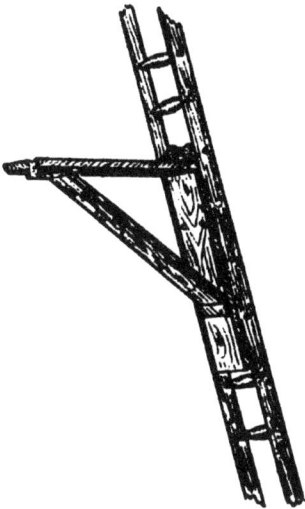
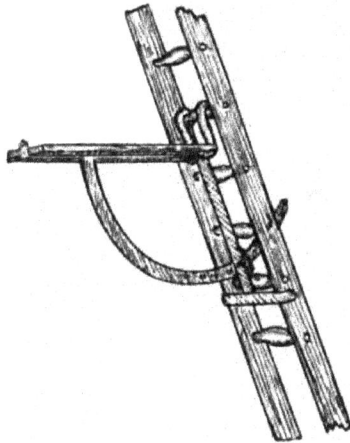

FIG. 66   FIG. 67

slope of ladder. The bracket should be long enough to carry a platform three boards wide, but as a rule it carries two.

Cripples may project from either side of the ladder, and are usually hung on the rungs. An advantage is gained if, in addition to this, clips are provided to clutch the sides of the ladder.

Buckets and Skips.—Besides the ordinary pail, which needs no description, larger buckets are commonly used for carrying concrete, mortar, earth, or any other moist or friable material.

Fig. 68 shows the tipping bucket, or skip, which balances on its hinges at A. The hinges are so placed that they are above the centre of gravity of the bucket when empty, and below the centre of gravity when full. This position allows the bucket to remain upright when empty, but it will make half a revolution and empty its contents when full. To prevent this action occurring before it is required, a catch on hinges is fixed on the rim of the bucket at B.

While the catch is in the position shown, the bucket cannot tilt, but if it is turned back the bucket makes the half revolution required, and after emptying its contents, swings upright of its own accord.

Buckets are constructed of steel, and the standard sizes vary in capacity from $\frac{1}{4}$ to 1 cubic yard.

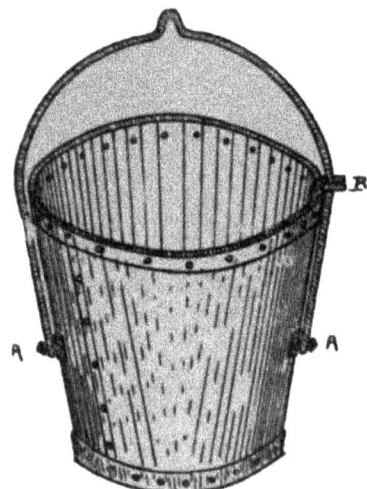

FIG. 68

For a similar purpose a steel box is used. In this case the bottom of the box is hinged, and on the catch being released, drops out, allowing the material to fall over any desired spot. The catch can be released from above or below by means of a chain connected thereto, and the bottom of the box regains its position when lowered to the ground for refilling.

Each box is fitted with a bow for chain hook, or lugs for chain slings ; has a capacity of about 3 cubic feet, is made of steel plates, and may be round or square on plan.

Baskets.—Baskets (as shown in figs. 69–73) have a capacity of about 1 cubic foot.

There are three qualities of cane used in their construction: 'Mackerel back,' recognised by its peculiar markings, 'Short Nature,' and 'Squeaky.' Of these, the first is the best, the others following in the order named. It is a defect of the baskets as ordinarily constructed that their handles and bottoms give way after very little

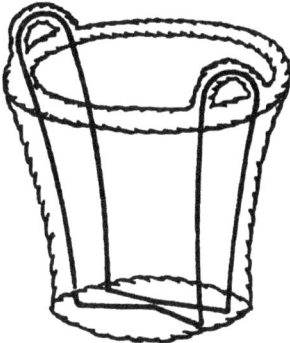 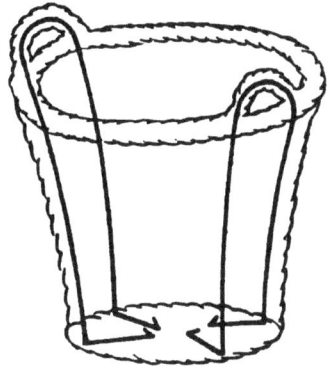

FIG. 69   FIG. 70

wear. Several improvements have been put on the market, the best of which are shown as follows.

In fig. 69 the black line represents an iron hook bent to the shape required, and the cane plaited round as for the ordinary basket.

It is claimed that the handles and bottoms of these baskets cannot give way, and it is a claim that is probably correct.

Owing to the difficulties of construction due to the rigidity of the iron hoop, they cost more than the ordinary basket, and this, with their extra weight, is unfortunately against their general adoption. Variations of the same idea are shown on figs. 70 and 71.

FIG. 71

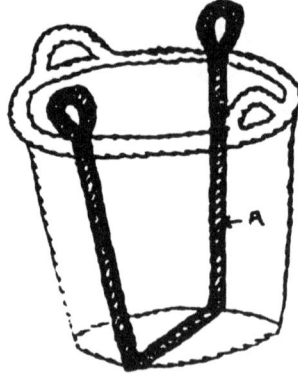

FIG. 72

In the first case (fig. 70) the iron is in two parts, which theoretically would allow of weakness, but in practice the basket answers its purpose well.

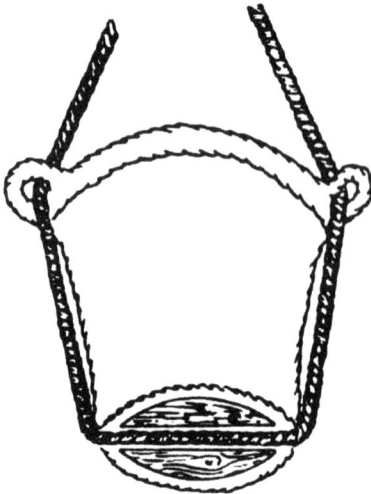

FIG. 73

In fig. 71 the rigid ironwork is placed by a wire rope spliced to make a complete circle. This kind of basket is easier to make and less in weight than those just mentioned, but the cost of the rope keeps the price high.

Fig. 72 shows another safety arrangement. A is a tarred hemp rope built into the basket as shown, and the ends fitted with eyelets for hoisting purposes, the handles being kept for use ███████ ██████kmen.

The arrangement is a practical one, and gives the required element of safety to the baskets so long as the rope remains sound.

Ordinarily constructed baskets can be made temporarily safe by passing the slinging rope or chain through the handles and round the bottom of the basket, as shown on fig. 73. To prevent the rope slipping, and to give the basket a flat bottom, pieces of wood can be fitted as shown.

Navvy Barrows.—Navvy barrows (fig. 74) are of hard wood, wrought and cast iron fittings and steel axles. They are fitted with iron, or wooden wheels bound with iron, and vary in weight from 60 to 75 lbs., and have a capacity of about $\frac{1}{10}$ of a cubic yard.

A barrow of this class can be slung by passing a hook through the wheel and rings round the handles.

Stone Bogies.—Stone bogies (fig. 75) can be fitted with plain wheels for running on flat surfaces, or flanged wheels for rails. They are of oak, with steel axles and cast-iron wheels. The handles for pulling are detachable and adjustable to either end.

Hand Barrows.—Hand barrows as fig. 76 are useful for carrying light loads, and, when bearing material that cannot roll, may also be slung.

Hods.—Hods (fig. 77) are used on small jobs in which to carry mortar, bricks, &c. In capacity they will hold $\frac{2}{3}$ of a cubic foot of mortar or twenty bricks, but an ordinary load is 16 walling or 12 facing bricks, the weight of which is considered to be enough for a man to carry up a ladder.

H

FIG. 74

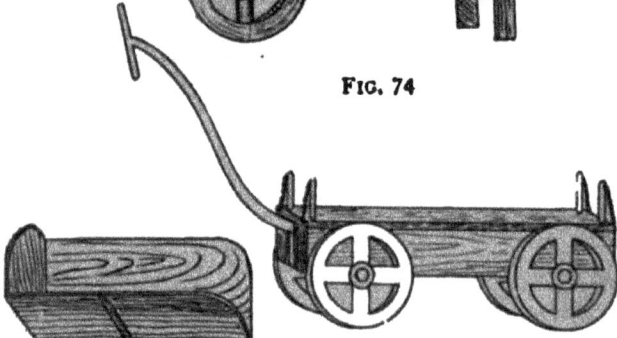

FIG. 75

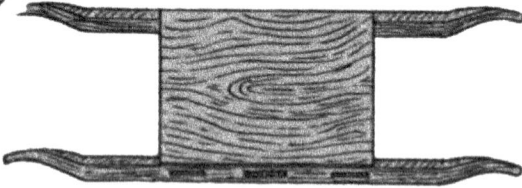

FIG. 76

FIG. 77

FIG 78

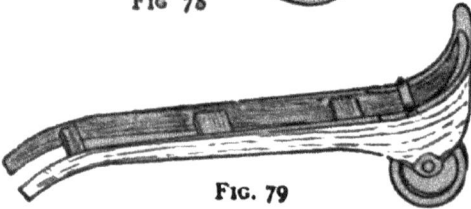

FIG. 79

Timber Trucks.—Timber trucks (fig. 78) are used for carrying timber balks, iron girders, &c. They are usually 3 or 4 feet in length, with a width of about 24 inches and a height of 22 inches. They are made sufficiently strong to carry 6,000 lbs.

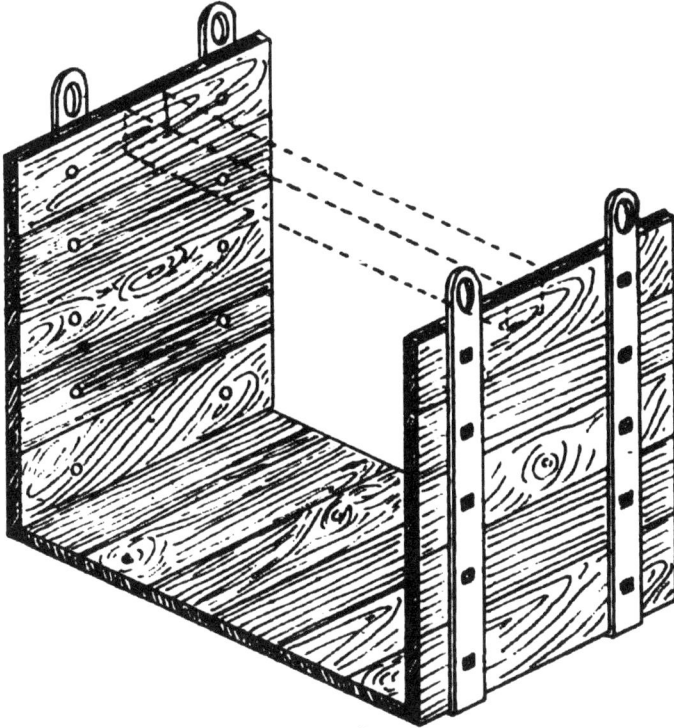

FIG. 80

Sack Trucks.—Sack trucks (fig. 79) are constructed of hard wood, with fittings of wrought and cast iron and steel axles. They vary in length up to 4 feet 4 inches, and the foot iron projects from 6 to 9 inches.

Crates, as shown on fig. 80, are constructed of oak with iron bindings. They will carry a weight of 1,500 lbs. and hold 350 bricks. They can be filled in the

303239   H 2

builder's yard and transferred direct to the working platform without disturbing the material, which, for saving time, is often of great advantage. The absence of sides facilitates loading, but on the other hand, if any materials, say bricks, are put in loosely, they may fall out during transference, causing danger to the workmen.

When used to carry rubble work which cannot be stacked, it is better that sides should be fitted.

When used to carry a roll of lead, a stay should be placed, as shown by dotted line on figure. This will prevent the crate buckling at the bottom.

These crates are sometimes fitted with wheels to run on rails.

Ashlar Shears.—The shears (figs. 81 and 82) are useful for lifting dressed work, the points fitting into

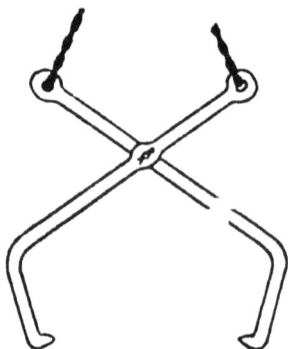 

Fig. 81   Fig. 82

small holes which have been cut out for their reception in the ends or sides of the stone. There is danger in their use if the points drag upwards and outwards. To prevent this as far as possible, the holes should be cut low, but not below the centre of gravity of the stone, or else it would turn over and perhaps fall.

Fig. 82 is a bad form of shears, as, owing to the sharp curve, the points can only clutch near the top of the stone.

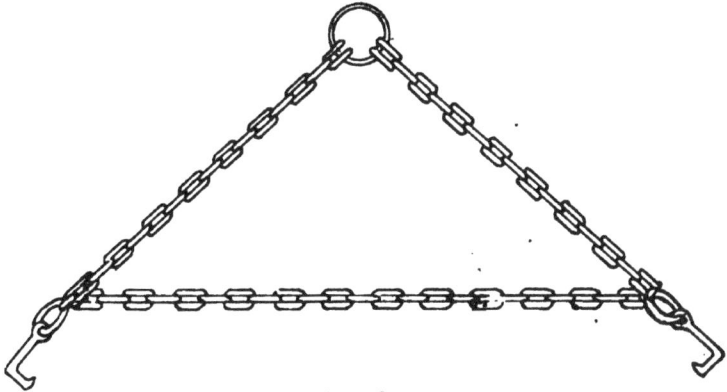

FIG. 83

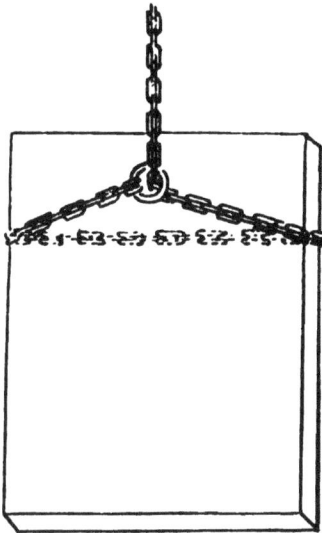

FIG. 84

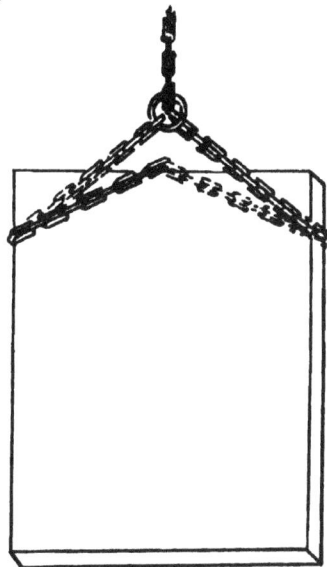

FIG. 85

Stone Clips and Slings.—The clips (fig. 83) are useful for lifting stone slabs. The hook rings slide along the chain, and the clips are therefore adjustable to any stone not exceeding in width half the total length of the chain.

The chain slings have a ring at one end and a hook at the other, and are useful for a similar purpose ; but the manner of slinging depends upon the thickness of the stone. For instance fig. 84, known as jack slinging, answers well with a slab, say, of over 6 inches in depth, but a thinner slab lifted in this way would be liable to break in the middle. If, however, the chain were placed as fig. 85, and which is known as figure-eight slinging, this risk would be removed.

Stone Lewises.—Lewises may be divided into two classes, curved and straight-sided.

Fig. 86 shows the first, and fig. 87 the second class.

FIG. 86

FIG. 87

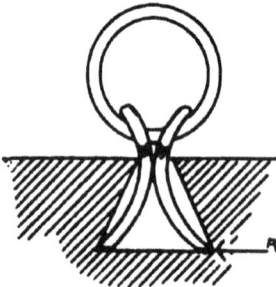
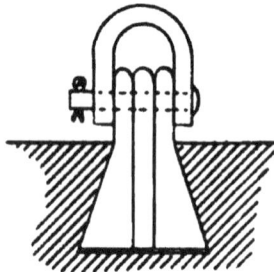

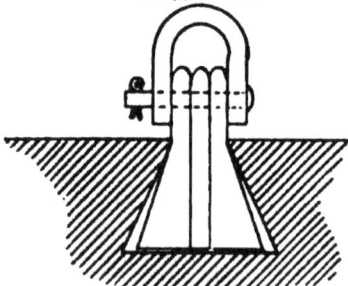
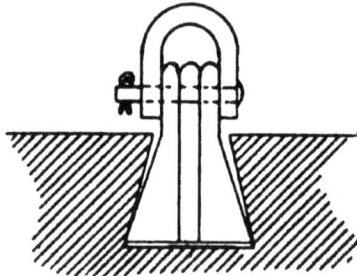

FIG. 88

FIG. 89

The first class is the inferior, as, when fitted into the stone, any jerk of the supporting chain would act at the points A as a blow on the stone, thus increasing any tendency to fracture.

The hole for the reception of the lewis is cut, so that a line down its centre would run across the centre of gravity of the stone ; and it is made as deep as may be required by the weight and hardness of the material.

The side or splayed pieces of the lewis shown on fig. 87 are fitted first, and the centre piece last. A bolt through the top fixes their position and also the ring by which it is to be lifted.   ·

Care should be taken that the sides of the second class of lewis fit accurately, for if they fit as fig. 88 they may flush the edge and break out, or if they fit as fig. 89 the risk of fracture, as in the first class, presents itself. In any case there is always a danger of mishaps occurring, especially where the stone is not free from vents.

Their use with safety can only be left to the judgment of the mason.

Stone Cramps.—The cramps tighten on the stone by means of a screw thread, as shown on fig. 90.

They are useful for lifting light finished work. Packing should be placed at AA to prevent damage.

FIG. 90

The ring by which it is slung is movable to preserve equilibrium.

Wire and Chain Scaffold Lashings.—Wire rope scaffold lashings are now to be obtained for use in place of fibre cords. They are made in lengths from 12 to 18

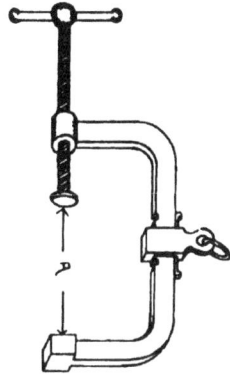

feet, and are **fitted at one end with an eyelet.** **In fixing,** they commence **with a clove hitch, the knot being con-** tinued as **with a fibre cord until near the end, when the** lash is taken **through the eyelet (see fig. 91) and finished** with jamming **turns.**

    It is **claimed that** no **wedges are required for** tightening wire **rope lashings, as they do not shrink** or swell ; on the **other hand, owing to their small circum-**

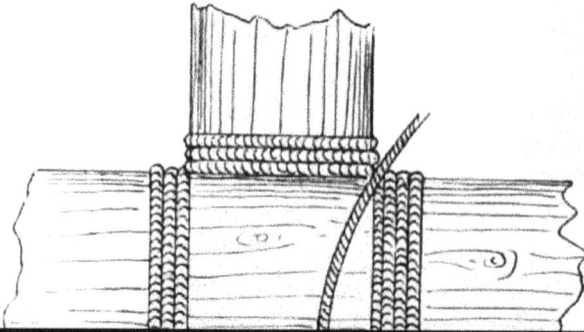

to standards is shown on fig. 92. It is easily and rapidly adjusted, and is tightened by means of screw nuts at A and B.

Permanent injury might, however, be done to the standards by the cutting in of the brackets when screwed up, especially after regular use. The possible loss of the parts and their weight and consequent disadvantage in transport are against their general adoption.

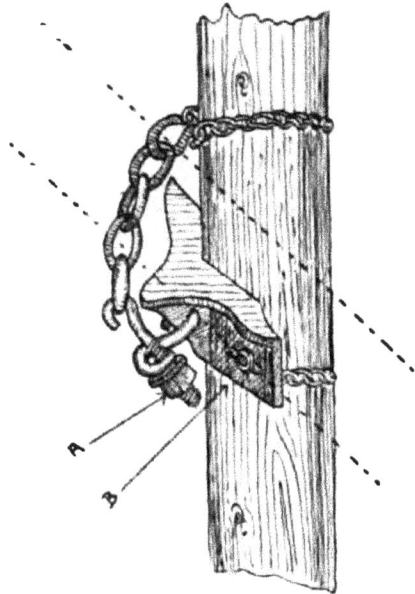

FIG. 92

Tightening Screws.— Tightening screws or coupling links (fig. 93) are fixed in the length of chain that connects the guys of the Scotch cranes to the base of the queen legs.

Under the continuous vibration of the scaffold they run down and release the chain considerably.

This can be pre-

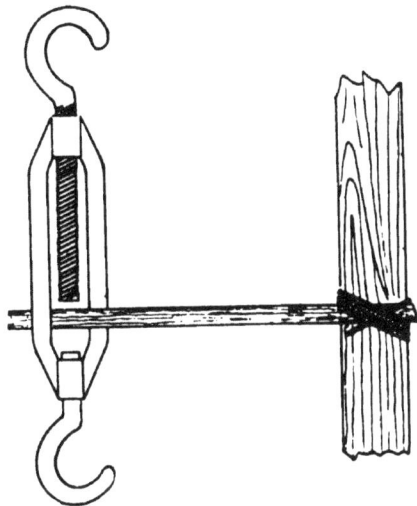

FIG. 93

vented to some e
shown on fig. 93,
member of the l

FIG. 9

moving heavy mat
l'egs should b
fig. 94, to form a l

Dog Irons.—Dog irons (fig. 96) are bars of flat or round wrought iron, turned up at the ends, which are pointed. If both ends point in the same plane they are

FIG. 96

termed 'male, if otherwise 'female.' The shank is about 12 inches long. Besides holding the timbers together, they exert a certain power of compression upon the joint they en-close. This is gained by hammering the inside of the spikes to a splay, leaving the outside to form a right angle with the shank.

FIG. 97

They may be described as inferior straps, and their holding power is from 600 to 900 lbs. per inch in length of spikes, as deduced from experiments by Captain

FIG. 98

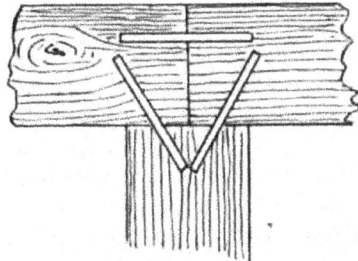

FIG. 99

Fraser, R.E. Dog irons have the advantage that their use does not injure the timber to any extent, and so depreciate its value. Dogs are fixed according to the joint to be enclosed. If the joint is at right angles to the run of the timbers, they are fixed as fig. 97.

If the timbers are at right angles they are fixed as fig. 98.

If both these joints occur the irons are placed as fig. 99.

They should be fixed on both sides of the timbers joined.

Bolts.—Bolts (fig. 100) are of wrought iron, and their different parts should be in the following proportions :

Thickness of nut	= 1 diameter of bolt
„ head	= $\frac{3}{4}$ „ „
Diameter of head or nut over sides	= $1\frac{3}{4}$ „
Size of square washer for fir . .	= $3\frac{1}{4}$ „
„ „ „ „ oak .	= $2\frac{1}{4}$ „
Thickness of washer . . .	= $\frac{1}{4}$ „

There are disadvantages to the use of bolts in scaffolding. For instance, the beams are weakened by the cutting of the fibres ; and, if the timber shrinks, the bolts may become loose. On the other hand, they can be easily tightened after the framing has settled into position.

Their strength depends upon the quality of the iron, but varies between 20 and 25 tons of tensile strain per square inch of the smallest sectional area (Anderson).

Washers are used to prevent the nut sinking into the wood when tightened, and are equally necessary, but not always seen, under the head. They should not be cut into the under side of timbers subjected to a cross strain,

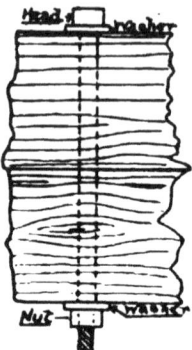

FIG. 100

as the cutting of any fibres is a source of weakness. Bolts are used where dogs and spikes are of insufficient length or holding power.

Straps.—Straps are wrought-iron bands of different designs, and are used to form a connection between timbers. Branched straps (fig. 101) are used to strengthen angle joints. They are usually fixed in pairs, and being fastened on the surface of the timbers they have an advantage over bolts in that they do not cut into the material. If the timbers settle at all, the straps may become subject to cross strains.

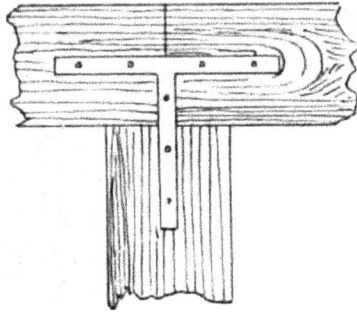

FIG. 101

Wire Ropes.—Wire ropes are now in general use for heavy purposes.

They are stranded and laid similarly to fibre ropes. They should be of mild plough steel wire. The number of wires in a strand varies from 12 to 37, and the number of strands is usually 6.

The following table gives the breaking strains of the ropes according to their circumference, and the least diameter of barrel and sheaves around which they may be worked at slow speeds.

In the table (p. 110) the diameters of the pulleys, &c. may be slightly reduced for the more flexible ropes, but better results can always be gained by using pulleys and sheaves of larger diameters.

A few points on the working of these ropes may be useful.

To remove a kink throw a turn out; it cannot be taken out by strain.

The ropes should be ungalvanised, and kept greased with any oil that does not contain acid or alkali.

A rope running in a V groove has a short life.

A rope that is allowed to ride, chafe on its own part or to overlap, will be almost immediately crippled.

The sign of an overloaded rope is excessive stretching.

Flexible Rope. 6 strands, each 12 wires			Extra Flexible Rope. 6 strands, each 24 wires	Special Extra Flexible Rope. 6 strands, each 37 wires
Size Circum.	Diam. of barrel or sheave round which it may be worked at a slow speed	Guaranteed Breaking Strain	Guaranteed Breaking Strain	Guaranteed Breaking Strain
Inches	Inches	Tons	Tons	Tons
$1\frac{1}{2}$	9	4	$7\frac{1}{2}$	8
$1\frac{3}{4}$	$10\frac{1}{2}$	$5\frac{1}{2}$	$9\frac{3}{4}$	11
2	12	7	13	$14\frac{1}{2}$
$2\frac{1}{4}$	$13\frac{1}{2}$	9	$16\frac{1}{4}$	$17\frac{1}{2}$
$2\frac{1}{2}$	15	12	$20\frac{1}{2}$	22
$2\frac{3}{4}$	$16\frac{1}{2}$	15	24	$26\frac{1}{2}$
3	18	18	$28\frac{1}{2}$	$32\frac{1}{4}$
$3\frac{1}{4}$	$19\frac{1}{2}$	22	34	$37\frac{1}{2}$
$3\frac{1}{2}$	21	26	39	43
$3\frac{3}{4}$	$22\frac{1}{2}$	29	$45\frac{1}{2}$	50
4	24	33	$51\frac{1}{2}$	$56\frac{1}{2}$
$4\frac{1}{4}$	$25\frac{1}{2}$	36	59	65
$4\frac{1}{2}$	27	39	65	$70\frac{1}{2}$
$4\frac{3}{4}$			74	79
5			$82\frac{1}{2}$	88

(*Bullivant & Co. Ltd.*)

Chains.—The strength of a chain depends upon the diameter and quality of the iron of which the links are formed, governed by good workmanship. The safe load for working can be calculated approximately by the following method:—

Square the number of eighths of an inch which are contained in the diameter of the iron of which the link is made, and strike off the last figure as a decimal.

For example, where the iron is of $\frac{1}{2}$-inch diameter, square the number of $\frac{1}{8}$ in the diameter, i.e. $4 \times 4 = 16 = 1\cdot6$ tons.

Generally before leaving the factory, chains are tested up to half the weight they should break under, and which is about double the load they are intended to carry in practice. This test cannot be relied upon for the future working of the chain, as any stretching of a link, which would ultimately result in fracture, would probably not be apparent under it. The links should therefore be examined periodically for any appearance of weakness or stretching.

A stretched link should at once be cut out, as it may break with much less load than that which it was first tested to carry. A chain during use also deteriorates in quality, and it is a good rule to have it periodically annealed.

The links should then be re-tested up to double the weight they are again required to carry.

A reliable chain is made of the treble best Staffordshire scrap iron.

Crane and pulley chains should be made with the shortest link possible, according to the diameter of the iron used, as there is a considerable leverage exerted on a long link when running round a pulley, more especially where the diameter of the pulley is small.

A Slater's Truss.—Slaters' trusses (fig. 102) are used in pairs by slaters and tilers when laying th - material. Boards are laid across the trusses and form a. effective platform on which the workman can kneel without damage to that part of the roof already covered. They are slung from the ridge or other suitable fixture, and can be pulled higher as the work proceeds. An old sack or similar material laid under the truss will prevent any possible damage during the progress of the work.

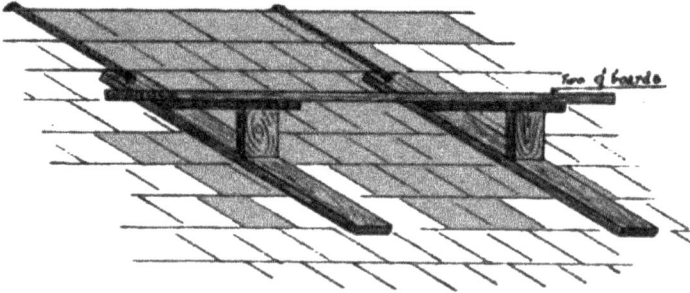

Two of boards

Duck Runs.—Duck runs (fig. 103) are laid upon slate and tile roofs to give footing to, and to prevent damage being done by, the workmen.

They should be firmly fixed, either by slinging from the ridge or butting against a solid resistance.

Fig. 103

Mortar Boards.—A mortar board is used as a bed on which mortar can be mixed or deposited. It is roughly made of four or five 9-inch boards each 3 or 4 feet long, framed together on the under side. Their use prevents the new mortar coming in contact with the scaffold boards with an injurious effect.

Wedges.—A wedge is a movable, double-inclined plane, used for separating bodies, and by this means, tightening any connections between the bodies they tend to separate. For scaffolding purposes they should be of oak, or other wood which gives considerable resistance to pressure across the grain. For tightening cordage wedges should be about 12 inches in length and, as far as possible, split to shape. In cross section they should be semicircular. Their taper should be gradual and not too sudden, as otherwise they might work out. When used in pairs as for shoring purposes, they are rectangular in cross section, and are termed folding wedges.

Nails.—Cut nails stamped out of plates are best for scaffolds. These nails have the advantage of being easily drawn out of timber. When driven with their flat sides the way of the grain, they do not tend to split the wood. They are used to fix platform boards, and sometimes guard boards, on edge.

Spikes.—Spikes are nails above 4 inches in length. They form a cheap method of fixing. Captain Fraser, R.E., has computed from experiments that their holding power in fir is from 460 to 700 pounds per inch of length. the depth of cover plate being deducted.

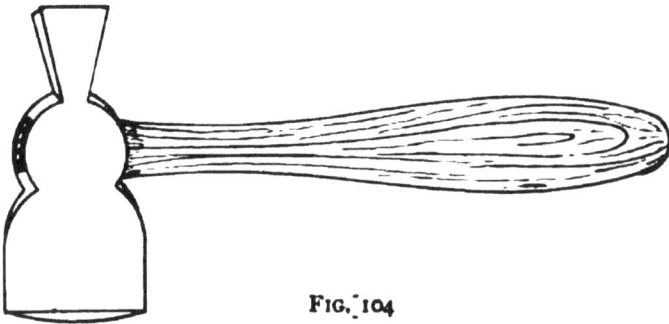

FIG. 104

Scaffolder's hatchet.—The scaffolder's hatchet (fig. 104) is an ordinary shingling hatchet with a hammer

I

head.　It is p
folder.　With
drive wedges,
round the mid
it as a lever.

THE transportation of material is not altogether within the province of a scaffolder, but it is so intimately connected—indeed, it is difficult to say where his connection with the lifting and carrying of material commences and finishes—that the subject is here briefly commented upon.

Crane Engines.—The engines of the crane are so arranged that all motions in connection with the derrick are under the control of the driver. The engines are double cylinder with link-motion reversing gear. The gearing is single and double purchase for lifting ; the jib barrel is fitted with steel catch wheel and double-lock safety catch to prevent the jib running down. The slewing gear is worked from the crank shaft, connected to the upright shaft from bottom race or spur wheel, and is wrought by worm and worm wheel with double-cone friction slewing gear. This arrangement permits of slewing the crane in either direction without reversing the engine. It might also be mentioned that the clutch for the jib motion is hooped with malleable iron to prevent the possibility of its bursting.

Crane engines can be worked by electrical, steam, or manual power. The smaller cranes are now so made that either steam or manual power can be used as required. It is of recent date that these engines have been supplied with electrical power, and of course their use is restricted to where this power is obtainable.

The Crane.—The crane consists of four parts, the mast, jib, sleepers, and guys or stays. The mast, which rises vertically, is connected at its base to the platform on which the engine stands ; and at the top, to the guys by a pivot which allows of rotation in a horizontal plane. It may be of iron, balk timber of oak or pitch pine, or in two pieces of the same, strutted and braced. The jib, which may be built of the same materials as the mast, is fastened to the lower end of the mast by a joint which allows of rotary movement in a vertical plane. The steel rope or chain which supports the weight runs from the drum placed near the engine and over the top of the mast and jib. Wheels are placed at these points to lessen friction.

The combination of movement allowed by the pivot of the mast and joints of the jib, enables the load to be carried to any point commanded by the effective length of the jib, except that it cannot be placed behind the guys. Jibs are used up to 70 feet in length. To prevent slewing under wind pressure, jibs over 50 feet long should be fitted with wind brakes, especially on exposed situations.

The crane will stand the greatest strain when the jib is most upright, and, reversely, less strain as it approaches the horizontal. It is a good rule, and one which works for safety, not to allow the top end of the jib to reach a lower level than the top of the mast, whatever the weight of the load carried may be.

Cranes are made suitable for derrick staging to carry a weight of 7 tons. If the boiler is attached to the rotating platform of the crane, it helps to counterbalance the load.

Cranes, while offering the readiest means of dealing with heavy weights, do not give the best results when used for placing material in its final position on the building. The vibration of the engine, the swaying of the supporting rope from the jib, and the unevenness of lowering under the band brake, prevent that steadiness of the material which is necessary for good fixing.

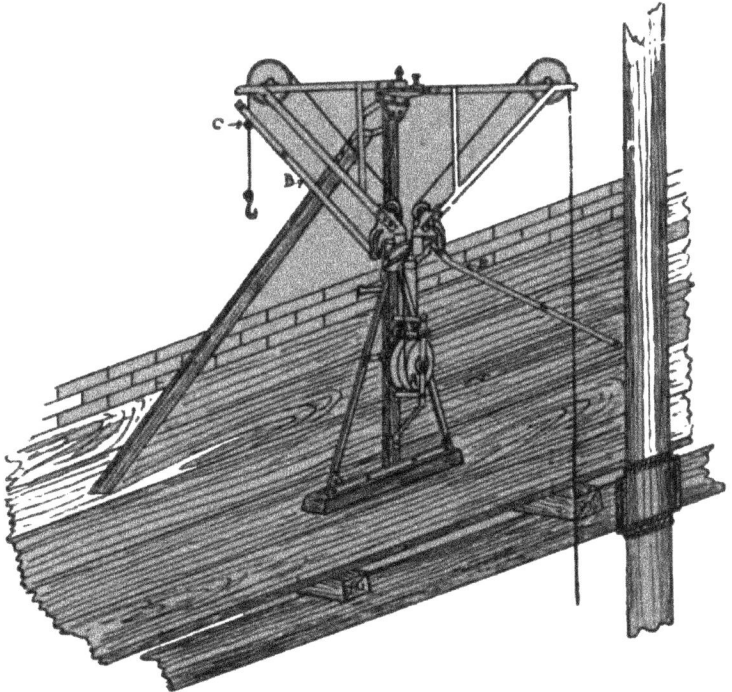

FIG. 105

Fig. 105 shows a small building crane ; it is worked by manual power, and is very suitable for light work. The illustration shows the general method of construction, but there are other patterns which give greater power.

The crane is fitted with two hoisting ropes which are wound on the drum at A. One rope rises while the other descends.   The ropes pass through the arms B, and when the catch C rises against the slot, it lifts the arm up.   The base of the jib to which it is connected then rises in the sliding groove and swings inward, carrying the load well over the platform where it is to be deposited.   When the new load begins to rise, the jib swings outward and downward, the rope paying out as required.   By this means the jibs are in use alternately for lifting.

Pulleys.—The pulley (fig. 106) is a circular iron disc which revolves freely on an axle fitted into an iron box. The circumferential edge is grooved to receive the rope or chain which passes round it.

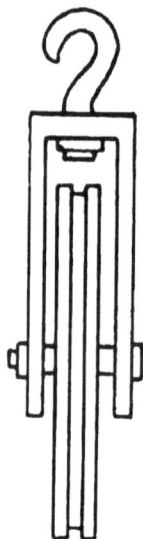

Fig. 107 is the section of a groove suitable for fibre rope driving.   The rope is gripped at its sides, thus increasing its driving power.

Fig. 108 is the section of a groove where the pulley is used as a guide only, the rope being allowed to rest on the bottom.

Fig. 109 is the section of a groove used for wire rope.   The groove is lined on the bottom with pieces of wood or leather to give greater friction, as the rope would be injured if it were gripped in a groove as fig. 107.

FIG. 106

Figs. 110 and 111 are the sections of grooves suitable for chains, the groove receiving every other link, the alternate links lying flat.

Fig. 110 is suitable where the pulley is used as a guide only, and fig. 111 is used for driving pulleys as in Weston's blocks.

The part elevation shown in fig. 112 is known as a

sprocket wheel, and shows the sprockets cast in the groove upon which the links catch. It is used for driving purposes.

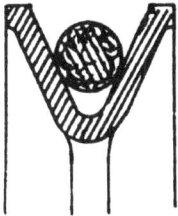
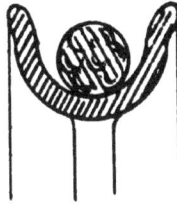

FIG. 107   FIG. 108   FIG. 109

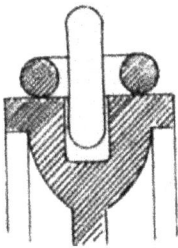
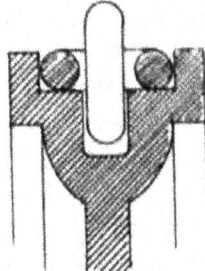

FIG. 110   FIG. 111

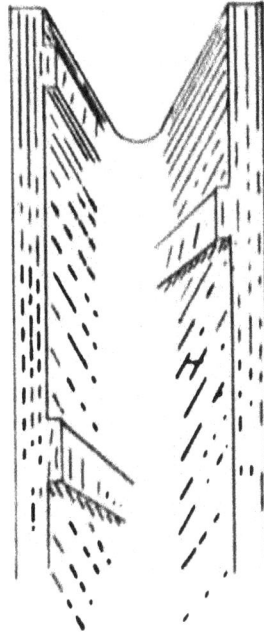

When the pulley cannot rise or fall, it is termed a fixed pulley, otherwise it is considered movable.

The fixed pulley, sometimes known as a gin wheel, can only change the direction of a force, and gives no mechanical advantage, but when used in conjunction with a movable pulley a mechanical advantage is gained.

FIG. 112

Fig. 113 is an illustration of the single movable pulley. The rope is connected to the beam at A, and passes round the pulley B and over the fixed pulley C

Now, if power be
sufficient to move
every one inch w
fore the mechani
workman pulling
lift w weighing
only half the spe

Various com
tions of pulleys
possible, but the 1
common in use
buildings is show
fig. 114, which i

Theoretically, there is no limit to the number of pulleys and consequent mechanical advantage, but the friction produced, and want of perfect flexibility in the rope, prevent any great increase in the number.

Differential Pulleys.—A differential block on Weston's principle (fig. 115) consists of a compound pulley of two different diameters but of one casting, and therefore rotating together. The chain is an endless one, and passes in turn over each diameter of the pulley. One of the loops thus formed carries a single movable pulley, while the other loop hangs loose (see fig. 116). The power which may be applied to the loose loop on the side which comes from the largest diameter will cause rotation of the pulley.

The chain must be four times in length the distance through which it is required to raise the load.

These pulleys are tested to 50 per cent. above the weight they will have to lift in practice, and the maximum load they will carry is stamped on the castings. The mechanical advantage derived depends upon the difference of diameter in the compound pulley. Usually with these machines two men are required to lift one ton.

Another common form of differential pulley is known as the worm block, and consists of two cast-iron toothed wheels at right angles to each other, connected by a worm thread of case-hardened, mild steel forging. The wheel upon which the power acts is worked by an endless chain, and the lift wheel may be fitted with a chain or wire rope to which the load is attached. Pulleys of this kind possess in a great degree steadiness in lifting or in lowering. This is due to the great mechanical advantage that can be gained by their method of construction. By these pulleys one man can

lift up to 3 tons.
maximum safe l
The friction betv

The Winch.—A winch is a hoisting machine in which an axle is turned by a crank handle, and a rope or chain wound round it so as to raise a weight. It is actually a form of lever whereby a weight may be moved through the distance required.

Fig. 117 gives a type of winch in its simplest form. The mechanical advantage gained by its use depends upon the difference between the radius of the driving wheel and the radius of the axle ; or the circumference of the wheel and the circumference of the axle.

If the radius of the axle were the same as the radius of the wheel, no mechanical advantage would be gained by its use. The advantage that is gained by the arrangement can be calculated as follows :

As the radius of the wheel is to the radius of the drum so is the weight that can be lifted to the power applied to the handle.

Example.

Let radius of wheel = R,
    radius of drum = r,
    power applied to handle = P,
and weight lifted = W.

$$\frac{R \times P}{r} = W.$$

Then if R = 12 inches,
    r = 4 inches,
    P = 60 lbs.

$$\frac{12 \times 60}{4} = \frac{720}{4} = 180 \text{ lbs. weight that can be}$$

lifted, giving a mechanical advantage of 3.

It is obvious that, owing to the restriction of size, the mechanical advantage that can be gained by the simple machine shown on this figure is limited. To overcome this difficulty one, two, or three pairs of toothed wheels are introduced into the machine, being thus known as a single, double, or treble purchase winch or crab. The difference in the number of teeth between the pinions and wheels gives the increased mechanical advantage that is required. The method by which to find the mechanical advantage gained is as follows :

The pressure exerted on the handle is to the weight lifted as the radius of the drum multiplied by the number of teeth in the pinions is to the radius of handle multiplied by the number of teeth in the wheels.

Winches, besides lifting from the barrel, are also used in conjunction with pulley wheels to change the direction of the force and to gain additional mechanical

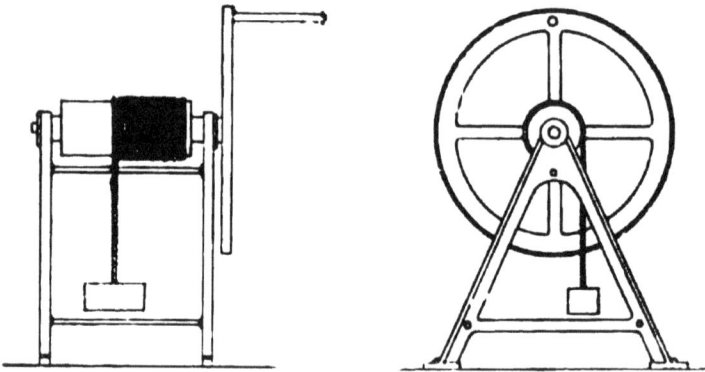

FIG. 117

advantage. If a double rope be used, considerable time will be saved in the progress of the work.

Fig. 118 shows the double rope. The middle of the rope is given a few turns round the drum, and the ends are continued over the pulleys, one sufficiently far to reach the ground. On raising the load the higher end of the rope descends, and is ready to carry up the next load by the time the first has been taken off. The illustration also shows the winch at work in connection with one movable pulley; but unless the material is to be raised to different heights, the same system of pulleys should be used on each rope.

An improved winch with an advantage over those ordinarily used has the drum grooved in three diameters, so that, with a minimum of trouble, a choice of mechanical advantages is gained. There is no need with these winches to pass the rope several times round the drum, for as the rope works in a groove, greater friction is set up; and the clutches provided to keep the rope in contact with the barrel for three quarters of its circumference, still further prevent any likelihood of slipping.

Jibs.—For the fixing of the gear a jib (fig. 119) is sufficient to carry a fixed pulley. A jib is a single pole attached horizontally to the standards or ledgers above the platform upon which it is

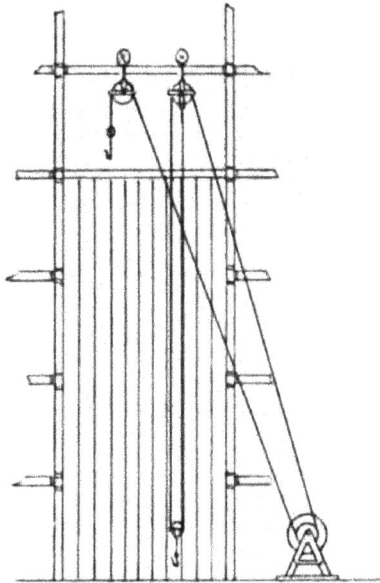

FIG. 118

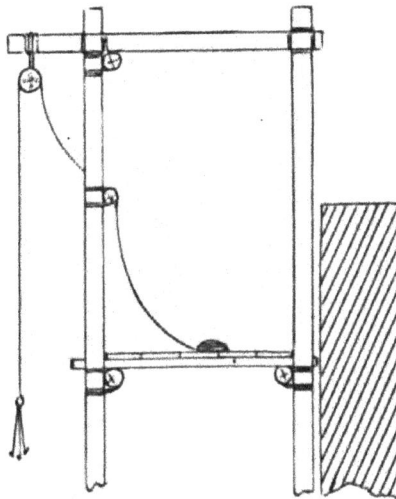

FIG. 119

intended to dc
slightly more t
hoisted, **usually**
It is **useful** a

This is arranged by carrying up on the interior of the building a series of standards and ledgers ; these rise from each floor as the work proceeds. The jib can be

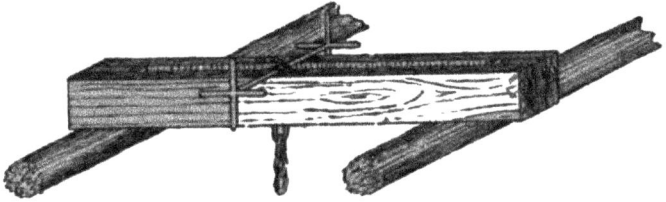

FIG. 121

carried right across the work in hand and the gear fixed as required.

Another form of jib known as the ' mason's ' is shown in fig. 121. It is of framed timber from 9 in. by 3 in. to 11 in. by 3 in., leaving a 4-inch opening down the centre, and rests across the ledgers. It allows the pulleys which are hung on to the iron movable axle, to be shifted horizontally throughout its length. For heavy material this is invaluable, as the load can be raised, moved to its position for fixing, and lowered as required.

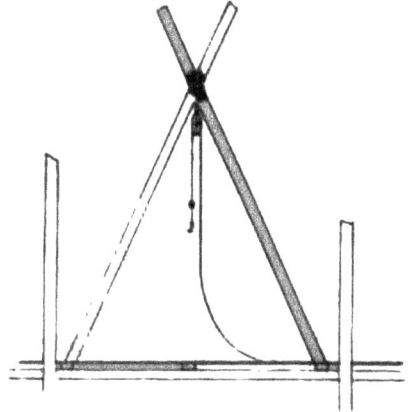

FIG. 122

Shears.—The shears or shear legs is another contrivance for supporting heavy weights. It consists of two legs forming two sides of a triangle, and may carry a pulley at the apex as shown in fig. 122, or a jib as shown in fig. 123. In the first case the legs are not rigidly fixed, but are kept in position by ropes, A and B, which, on being slackened, allow the shears to move from the perpendicular (fig. 124). In this manner loads can be lifted and placed in a different position other than that immediately over the one they first occupied. The range is, however, limited, as in practice the shears should not move more than 18 inches to 2 feet from the perpendicular.

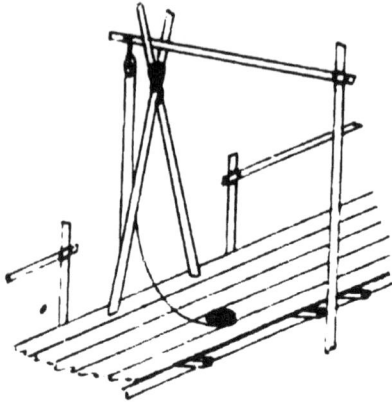

Shears are useful for raising and lowering the machinery on Scotch derricks, and are often fixed on gantries to carry jibs.

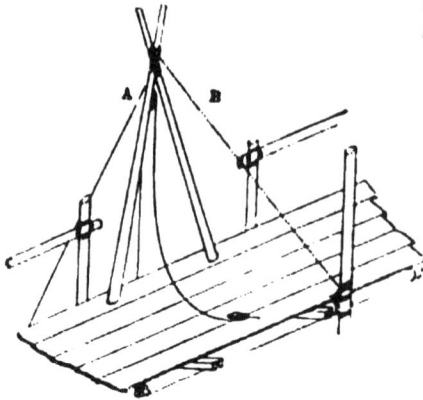

For heavy weights, the legs and also the jib can be of two or three poles tied together.

Fig. 123

Fig. 124

Gin.—The gyn, or gin, consists of three legs usually from 12 to 13 feet long. They are set up and joined together at the top, thus forming a triangulated frame. A pulley wheel or block is fastened at the apex, and extra power can be gained if they are furnished with a crab winch standing between two of the legs. They are useful in lifting or in lowering material through a well or opening in the working platform, as shown in fig. 125.

Rails.—Scaffolds of a particularly strong construction have, when necessary, rails laid upon them, in order that light trucks may be moved freely from place to place.

Sack trucks are also used on platforms to carry cement, &c., where required.

Other accessories for carrying purposes, the uses of which are obvious, are described in the chapter on Scaffolding Accessories.

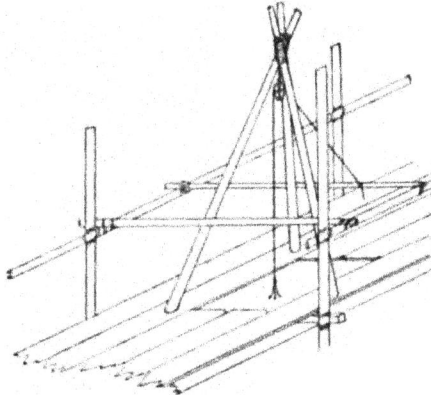

FIG. 125

The attachment of material to the transporting power is within the province of the scaffolder. To take each class of material separately :—

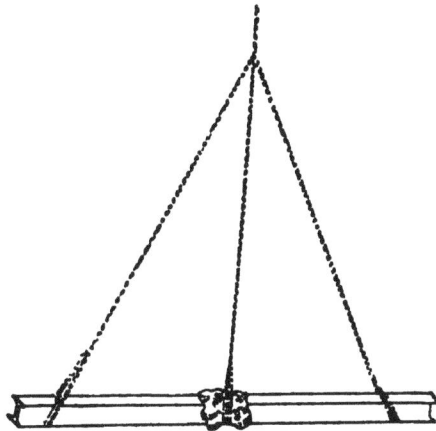

FIG. 126

Ironwork.—Ironwork is principally used in the form of girders and columns. These are sometimes slung by a chain round the middle, and as evenly balanced as possible. There is considerable danger of the chain slipping, however

K

well balanced ; more especially is this the case if the load is tilted when swinging. This may happen by the load receiving a jar through touching some part of the erection, and thus allowing the material to fall. To prevent this 'softeners'—i.e. old bags, sacks, or even pieces of wood—are placed between the chain and load. Then, with the chain turned twice round the whole tightly, the danger is minimised. An extra chain may also be run from each end of the load to a point some distance up the supporting chain, as shown on fig. 126.

Timber.—Timber in lengths can be carried in the same manner as ironwork, but, owing to the greater friction set up, it is not so likely to slip as the former. The same precautions should be taken.

To carry timber or ironwork vertically, the supporting chain is given a timber hitch round one end of the pole, and a half hitch round the end which is meant to rise first. It is sometimes advantageous to substitute a cord lashing for the half hitch. Then, when the highest end of the pole reaches the platform, the lashing can be removed and the pole received horizontally. This method is useful where the load has to be passed through a window.

Bricks, slates, &c., are slung in crates and baskets, and on small jobs are carried in hods by labourers. These accessories are described in Chapter VI.

Note should be taken that these fittings are in the first instance strongly made, kept in proper repair, not overloaded, and that spring hooks are used on the slings.

Stone.—Stone-work can be slung by means of the lewis, slings, cramps, clips, or shears. Another method is to pass the chain several times round the material, as for girder lifting. It is only suitable for rough work, as any finished edges or chamfers may be flushed even if 'softeners' are used.

A SCAFFOLD, considered as a whole, is in a stable condition when, under the forces that may act upon it, it remains in a state of rest or equilibrium. Two forces which tend to create a loss of equilibrium are: the pressure of wind which acts from any direction in a horizontal plane, and the force of gravity due to the weight of the scaffold and that of attendant loads.

Wind Pressures.—*The effect of wind upon a pole scaffold:*

The effect of wind acting on a single scaffold pole, erected as a standard, can be first considered. For this purpose the pole shall be taken as 32 feet long, 2 feet of which are below ground level. The force of wind depends upon its velocity, and it is measured by the pressure it exerts on a square foot of surface normal to its direction.

If a point in a body is fixed, so that the body cannot move out of its place, but may rotate about that point ; a force which acts at any other point, but in a direction that does not pass through the fixed point, will produce rotation.

In applying this principle to the effect of wind on the scaffold pole, the ground level will be the fixed point about which the standard may rotate.

The wind acting upon the exposed surface of the

K 2

pole may be likened to a series of parallel forces that, not acting through the fixed point, tend to produce rotation.

An advantage is gained if, instead of taking the wind as a series of parallel forces, it is considered as a resultant force of proportionate magnitude exerting a pressure upon the centre of the exposed surface. In practice it will be sufficiently correct to take the centre of surface of the pole at a point at half its height.

The tendency of a force to produce rotation about a fixed point is termed its moment about that point. It is measured by multiplying the units of force exerted by the units of the distance between its point of application and the fixed point.

Example: If at the centre of surface of the pole under consideration, that is at 15 feet above the fixed point, the resultant force of the wind is equal to a pressure of 100 lbs., the moment about the fixed point will be 15 by 100 = 1500.

In like manner the moment of resistance due to the weight of earth packed round that portion of the pole below ground can be estimated.

For the pole to remain in equilibrium it will be necessary for the moment of resistance to equal the moment of the overturning force, assuming that the fixed point is stable.

No practical good can result by pursuing this calculation further. It may be taken for granted that as the wind occasionally exerts a pressure of over 50 lbs. per square foot, and a pressure of 40 lbs. per square foot is the least for which calculations should be made, it will always be necessary when scaffolding to any height to adopt special measures to preserve stability.

When the ledgers are added to the standards they have some effect upon the equilibrium of the erection, and this must now be considered.

The wind can be taken as acting from two directions, first directly along the scaffold, and secondly across the scaffold. When at any other angle to the structure it will have effect in both of these directions, being greatest in the one with which it most nearly corresponds.

When blowing along the scaffold.—The standards and ledgers form with the ground level a series of rectangular parallelograms. The connections between the sides of the parallelograms are not rigid. They most nearly approach cup and ball joints, and as such, it will be wise to regard them as entirely loose to a rotating force.

The shape of a parallelogram with loose joints can be altered by a force acting in the plane of its surface, and this alteration of shape can take place without creating any strain on its joints or members.

From this it will be seen that the tyings and ledgers of the erection may be considered as offering no resistance to a force tending to rotate the standards about their fixed points ; but by adding to the surface upon which the force can act, the ledgers increase the overturning moment about the fixed point.

In practice, although the scaffold may not entirely fail, any change from regularity of structure due to wind pressure would cause the members of the erection to offer a less effective resistance to the other forces acting upon it. This being so, means must be taken to give that rigidity to the standards, without which the scaffold may collapse.

Although a parallelogram with loose joints will alter its shape under pressure, a triangle under similar conditions cannot do so.

Advantage is taken of this fact to obtain the rigidity which is necessary.

Taking a standard, and one of the ledgers, or the ground level as forming two sides of a triangle, a pole, termed a brace, is fixed to form the third side.

The triangle thus formed is a rigid figure offering resistance to any force acting upon it in the same plane as its surface, and it will remain rigid until the destruction of one of its joints or members. It follows, therefore, that the standard forming one side of the triangle becomes a sufficiently rigid body to withstand any pressure of wind that may act upon it. The other standards in the erection, if not tied to the brace, gain rigidity from the triangulated standard because of their connection thereto by the ledgers.

When the wind is blowing across the scaffold.—If the erection is of the dependent type, the standards, putlogs, wall of building, and ground level form a series of parallelograms which differ from those previously noted in that a sufficiently rigid angle is formed between the wall and ground level.

The effect of this inflexibility is to create rigidity throughout the parallelogram, always providing that the other sides are firmly connected at their points of juncture.

In practice this is not so ; the putlogs, if tied to the ledgers, which for this purpose is the same as being tied to the standards, have no fixed connection to the wall of the building ; but if they are supplemented by poles tied from the standards to within the building, they can be regarded as having, in effect, fixed joints.

If it be impossible to tie the standards within the building, the same effect can be gained by strutting from the ground level.

This scaffold, if so treated, is sufficiently rigid to withstand any wind force that tends to overturn the standard, either towards or from the building.

If the erection is of the independent type, the cross section also shows a series of parallelograms with loose joints, and so similar conditions exist as in the first example, except that the overturning force is acting in a different direction.

Any of the methods of gaining rigidity already given, and shown on figs. 21 and 24, can be applied in this instance.

Guard boards, rails, face boards, &c., have no other effect than that of increasing the surface upon which the wind can act. In consequence, the overturning moment of the standards about their fixed point is also greater.

Gantries form parallelograms with fixed joints with sufficient strength—unless carried to a great height—to withstand any wind pressure. If necessary, they can be braced in the same manner as the pole scaffold.

Scotch derricks are so strongly built that, unless a wind force exerting great pressure acted upon them, they could be considered safe from destruction by that means.

The four pillars standing square to form each leg are crossed at right angles by transoms which are bolted to the uprights. The parallelograms thus formed have joints which allow of rotation ; but the cross braces fitted in each bay give rigidity in two ways. Besides triangulating the frame, they offer a definite resistance to movement on the bolt by butting against the transom, as will be seen by reference to fig. 1.

This resistance to movement is to some extent due to the resistance of the timber to crushing.

The larger parallelograms formed by any two legs, the trussed beam and the ground level, have joints that can only be destroyed by very great force.

As the highest pressures noted in this country have equalled 80 lbs. per square foot, and therefore have to

be guarded again
shown on Frontis

The Force
of a body **equal**
drawn towards t)
The **weight** o
force acting verti

bottom ledgers.   The extra length of standards in this connection can safely be ignored.

If a body rests on a hard surface, it will stand or fall according as to whether a vertical line, drawn from the centre of gravity, falls within or without its base. The base of a body is within a line drawn round the points of support.

It will therefore be seen that so long as the scaffold is not acted upon by any other force it will remain in equilibrium.   But a scaffold is erected to carry weights, and the effect of these weights upon the stability must now be considered.   The effect of a load upon the scaffold is to alter the position of its centre of gravity.

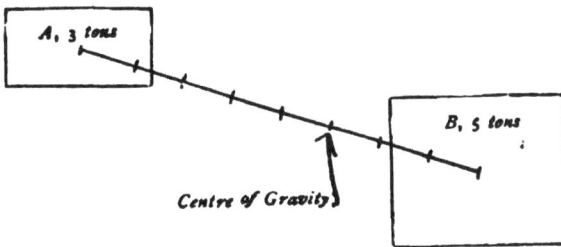

FIG. 128

Considering the scaffold and its load as two separate bodies, the point or centre of gravity about which the two combined weights would act is found as follows :—

Let A and B, fig. 128, represent two heavy bodies. Join the centre of A to centre of gravity of B ; divide this line into as many parts as there are units of weights in A and B together.   Then mark off from A the number of units there are in B, and the point thus found is the point about which the combined weights act, i.e. the required centre of gravity of the bodies.

The scaffold may, however, have several loads to carry, one or more of which may be materials slung on the hoist.

B divide off a n
the number of
thus found is t
process is conti
considered, care
in the bodies w
to the units of ٧

4. s

of gravity of the scaffold itself, would have no other effect than to slightly raise the centre of gravity of the structure.

To find the Centre of Gravity of Scaffold Boards laid to form a Platform.—A scaffolding platform, being of a slight depth in comparison with its length and breadth, may be treated as a surface usually rectangular.

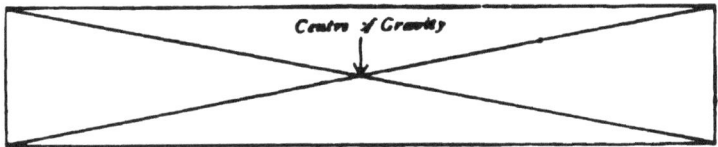

Centre of Gravity

FIG. 130

The centre of gravity of a rectangular surface is the point of intersection of its diagonals (fig. 130).

To find the Centre of Gravity of a Dependent Scaffold and the Effect of Loads upon it.—A dependent scaffold, having only one frame of standards and ledgers, to which are attached the putlogs, cannot be considered as an evenly disposed regular body. Nevertheless, the rule that the scaffold will not be in equilibrium unless a line from the centre of gravity fall within the base still holds good. In the case under consideration, as the wall of the building to which the scaffold is securely attached by the putlogs and ties, carries its share of the weight of the loads and putlogs, it must be taken as forming an integral portion of the scaffold itself.

Therefore the centre of gravity of a dependent scaffold will be the resultant centre of gravity of the outer frame, the putlogs, and the wall, considered as separate bodies.

The centre of gravity of the frame may be found

by taking it as a rectangular surface. If necessary, the boards may be treated in like manner.

The system of putlogs, if they are regularly placed, may be treated as a regular body, and the centre of gravity found by the method already given; but in practice their weight would have no effect towards loss of equilibrium.

The wall may also be treated in like manner if of even thickness. If of varying thickness, the centre of gravity of each portion of even thickness should be found.

The resultant will be found by the method already given for finding the centre of gravity of a number of bodies.

The base of the erection in this case should include the base of the wall.

The effect of ordinary loads upon the stability of a scaffold of this type is practically nil. No weight that the scaffold was capable of carrying in itself could bring the resultant centre of gravity of the scaffold and wall outside of the base; so that unless the scaffold failed from rupture of its members or connections, it may be considered safe from collapse due to instability.

To find the Centre of Gravity of a Gantry.—This can be found by the method given for independent pole scaffolds.

To find the Centre of Gravity of a Scotch Derrick.—Owing to the unevenly distributed weights about these scaffolds, they cannot be taken as regular bodies. It will therefore be necessary to take each part of the erection separately, and after finding the centre of gravity of each, to find the resultant centre of gravity of the mass by the method already given.

To find the Centre of Gravity of each Part.—Each leg can be treated as an evenly disposed rigid body.

The mass of brickwork that is placed at the foot of the legs may be treated similarly.

The platform may be considered as a surface. If triangular, the centre of gravity is found by the following method :—

Bisect the base and join the point of bisection to the opposite angle. The centre of gravity is at a point one-third of the length of the line measured from the side divided (fig. 131).

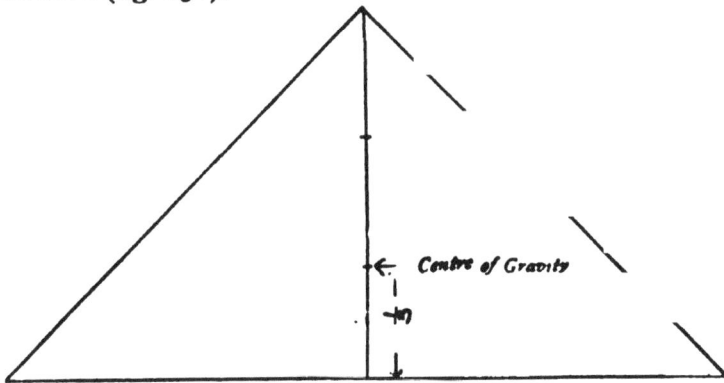

Centre of Gravity

FIG. 131

The centre of gravity of the joists supporting the platform can be taken with that of the platform itself, providing that the weights of each are added together and the centre of gravity of the platform only lowered to half the depth of the joists. The trussed girders supporting the platform may be treated as rectangular surfaces.

The centre of gravity of the guys, sleepers, and jib will be at a point in the centre of their length.

The centre of gravity of the engine may be somewhat difficult to find, but it will be sufficient to treat it as a cylinder. The centre of gravity of a cylinder is the middle of its axis.

Loads may be of various forms, but the centre of gravity will be found on a line drawn downwards through the load immediately under the supporting chain. In actual practice no load, the centre of gravity of which, considered separately, falls within the base of the scaffold which supports it, will cause instability. The greatest effect it can have is to bring the centre of gravity of the entire mass nearly to the edge of the base, so that a comparatively light load acting from without may cause loss of equilibrium.

It has been so far assumed that, owing to the use of braces, ties, struts, &c., the scaffolds considered have been rigid bodies, and where this is so the principles given hold good. In practice, however, owing to the lack of, or only partial use of, the members just mentioned, scaffolds are often more or less flexible bodies. Where this is the case, the lack of rigidity greatly increases the danger of collapse, as the timbers, through yielding by flexure to the loads that act upon them, allow such an alteration of the shape of the scaffold that the centre of gravity may be carried outside the base. Even where this does not occur, the racking movement allowed is dangerous, as the connections are strained and become loose, creating an element of risk that the otherwise careful scaffolder cannot altogether remove. For the scaffolder the lesson to be learned is— that, whether the force he is dealing with arises from the wind, loads, or a combination of both, he must triangulate—TRIANGULATE.

It will be necessary to know the weight of material in working out these problems. These have been given in the Appendix.

THE strength of a scaffold equals the resistance that its members and their connections can offer to the strains that act upon it. The timbers used may fail in various ways.

First.—As beams, they may fail to resist a cross strain.

Secondly.—As pillars and struts, they may fail to resist compression in the direction of their length.

Thirdly.—As ties and braces, they may fail to resist tension ; that is, a strain tending to pull a beam asunder by stretching.

The loads that act upon a scaffold may be dead, that is to say, they do not create a shock ; or live, which means they are not stationary, and may cause shock and vibration.

A live load causes nearly double the strain that a dead load produces. If, therefore, both are of equal weight, the timber under strain will carry as a live load only one half of what it would carry as a dead load. The breaking weight is the load that will cause fracture in the material. The safe load is the greatest weight that should be allowed in practice. It is in proportion to the breaking load, and that proportion is termed the factor of safety.

Experiments have been made to determine the resistance of timber to fracture under the various forces that act upon it.

The result of these experiments is expressed by a given number, termed a constant, which varies with the

different growths
which they are su
    The constants
of timber under a

Material

Spruce . .
Larch . .
Fir, Northern ⎫
 ,, Dantzic ⎬
 ,, Memel ⎭
 ,, Riga . .
Elm . .
Birch .
Ash, English
Oak, English
 ,, Baltic .

    The factor of

The working load on pillars should not be greater than one-tenth of the breaking weight; but if the pillars are used for temporary purposes, and are over 15 and under 30 diameters one-eighth, and under 15 diameters one-fifth may be taken as the factor of safety.

TABLE III.—*The Constants* (E) *for Breaking Weight under tensional stress.*

Material	E in lbs. per unit of 1 sq. inch	Authority
Spruce	3,360	
Larch	3,360	
Fir	3,360	Hurst
Elm	4,480	
Ash	4,480	
Oak	6,720	

The working load should not exceed one-fifth of the weight that would cause rupture.

Beams subject to a Transverse Strain.—The loads that act upon a beam may be concentrated, that is, acting at one point, or distributed, which means that the load is evenly placed over the entire length of the beam or a portion of the beam.

An evenly distributed load is considered to act at a point immediately below its centre of gravity.

If a beam carries several loads they are considered to act at a point immediately below the resultant centre of gravity of the whole.

Beams may be fixed, loose, or continuous. When fixed they are built into the structure; if loose they are supported only; and are continuous when they have more than two points of support in their length.

For practical purposes, a continuous beam may be considered one-half stronger than when supported at two points only.

L

The strength of a solid rectangular beam varies directly as its breadth, the square of its depth, and inversely as its length.

The formula for finding the breaking weight (B.W.) of a solid rectangular beam supported at each end and loaded at its centre is as follows :

$$\text{B.W.} = \frac{b\, d^2}{l} \times c$$

where b = breadth in inches ;

d^2 = depth multiplied by itself in inches ;

l = length of beam between supports in feet ;

c = the constant found by experiment on the timber in use.

Example.—Find the B.W. of a solid rectangular beam of Northern fir, 9 in. by 9 in., and 10 feet between supports, loaded at its centre.

The constant for Northern fir (Table I.) is 448.

$$\text{B.W.} = \frac{9 \times 9^2}{10} \times 448 = \frac{9 \times 9 \times 9 \times 448}{10} = \overset{\text{tons}}{14} \quad \overset{\text{cwt.}}{11} \quad \overset{\text{qrs.}}{2} \quad \overset{\text{lbs.}}{11\tfrac{1}{2}}$$

One-fifth of this must be taken as a safe load, say 2 tons 18 cwt.

The strength of a solid cylindrical beam varies directly as its diameter cubed, and inversely as its length.

The formula for finding the breaking weight (B.W.) of a solid cylindrical beam supported at each end and loaded at its centre is as follows :

$$\text{B.W.} = \frac{d^3}{l} \times c \times \frac{10}{17}$$

where a^3 = least diameter in inches cubed ;

l = length of beam between supports in feet ;

c = the constant found by experiment on rectangular beams ;

$\dfrac{10}{17}$ = the proportion that cylindrical beams bear to square beams.

Example.—Find the B.W. of a solid cylindrical beam of spruce fir 6 inches in diameter and 4 feet between supports.

The constant C for spruce fir is (Table I.) 403.

$$\text{B.W.} = \frac{6 \times 6 \times 6 \times 403 \times 10}{4 \times 17} = 12{,}801\tfrac{3}{17} \text{ lbs.} = \overset{\text{tons}}{5} \quad \overset{\text{cwt.}}{14} \quad \overset{\text{qr.}}{1} \quad \overset{\text{lbs.}}{5\tfrac{3}{17}}$$

One-fifth of this must be taken as a safe load, say 1 ton 2 cwt. 3 qrs.

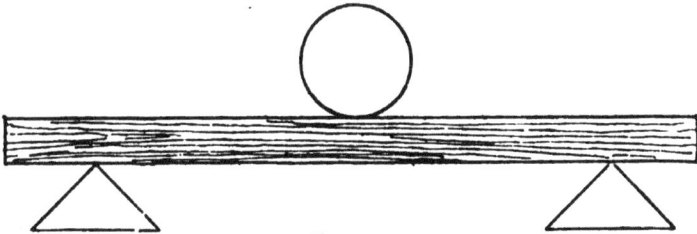

FIG. 132

The above diagram (fig. 132) illustrates the supports and loading of the beams just considered. The following diagrams show beams variously supported and loaded, and their relative strength to the first. By adding to the formulæ already given, the proportion that these following bear to the first, the same formulæ can be used for all.

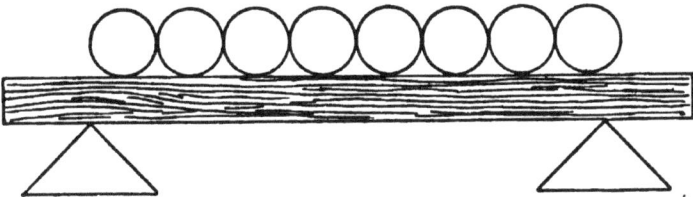

FIG. 133

Beam supported at both ends and load equally distributed bears double of fig. 132.

L 2

Beam fixed, load

fig. 132.

Beam fixed, load

fig. 132.

To find the effective length of a beam supported at both ends when the position of the load is varied :

Rule.—Divide four times the product of the distances of the load from both supports by the whole span, in feet.

Example.—Find the effective length of a beam the supports of which are 6 feet apart, the load acting at a point 1 foot from its centre.

The distances of the load from each support in this case are 2 feet and 4 feet.

Therefore $\frac{2 \times 4 \times 4}{6} = 5\frac{1}{3}$ feet, effective length of beam.

Posts and Struts subject to Compression.[1]—Posts and struts, when above 30 diameters in length, tend to fail by bending and subsequent cross breaking. When below 30 and above 5 diameters in length, their weakness may partly be in their bending and partly by crushing, and when below 5 diameters in length by crushing alone ; but, as these latter are rarely met with in scaffolding, they need not be considered.

On the resistance of long posts.—To find the greatest weight, W, that a square post of 30 diameters and upwards will carry :

Multiply the fourth power of the side of the post in inches by the value of D (Table II.), and divide by the square of the length in feet. The quotient will be the weight required in pounds.

The formula is as follows :

$$W = \frac{d^4 \times D}{L^2} = \text{weight in lbs. for square posts to resist bending,}$$

where d = depth in inches,
  D = constant in pounds, for flexure,
  L = length in feet.

---

[1] Tredgold's *Carpentry*.

Example.—Find the weight that may be placed upon a post of elm 12 feet long and 4 inches square.

The value of D (Table II.) being 1,620,

$$\frac{4' \times 1,620}{12} = \frac{256 \times 1,620}{144} = 2,880 \text{ lbs.}$$

One-tenth of this must be taken as a safe working load $= 288$ lbs.

The strength of cylindrical posts is to square ones as 10 is to 17, the rest of the formula remaining the same, except that $d =$ diameter of post in inches.

When the pillar is rectangular.—Multiply the greater side by the cube of the lesser in inches, and divide by the square of the length in feet. The quotient multiplied by the value of D (Table II.) will give the weight as by bending, the resistance to crushing is a considerable proportion of the strength. This must, in consequence, be allowed for in any calculation for the breaking weight.

To find the weight that would break a square or rectangular post of between 5 and 30 diameters in length :

Multiply the area of the cross section of the post in square inches by the weight in pounds that would crush a short prism of 1 inch square (Table II.), and divide the product by 1·1 added to the square of the length in feet, divided by 2·9 times the square of the least thickness in inches.

The formula is as follows :

$$W = \frac{D\,S}{1\cdot1 + \frac{L^2}{T^2\,2\cdot9}}$$

where D = the constant for the resistance to crushing.

S = the area of the cross section of the post in square inches.

1·1 is a modification introduced in order that the result may be in accord with the result of experiments.

L = the length of the post in feet.

T = the least thickness in inches.

2·9 = the constant for the resistance to bending, and which is taken at that figure for all timbers (Table II.).

Example.—Find the breaking weight of a post of Dantzic oak 10 feet long and 6 inches square.

The constant for Dantzic oak (Table II.) is 7,731.

$$W = \frac{7,731 \times 36}{1\cdot1 + \frac{100}{2\cdot9 \times 36}} = \frac{278,316}{2\cdot05} = 135,763 \text{ lbs.}$$

The factor safety is $\frac{1}{8}$. Therefore the safe load will be 16,970 lbs.

The strength of cylindrical pillars is to square ones as 10 is to 17.

Braces and Ties subject to a Tensional Strain.—
The weight that will produce fracture in a beam strained
in the direction of its length, is in proportion to the area
of the cross section of the beam multiplied by the
weight that would fracture a unit of that area.

The formula is as follows :

$$B.W. = E S$$

where E = the cohesive force in lbs. per unit of 1 square inch, as
in Table III.

S = the sectional area of the beam in square inches.

Example.—Find the B.W. of a rectangular elm brace
9 in. by 3 in. under a tensional strain.

The value of E (Table III.) is 4,480.

$$B.W. = 4,480 \times 9 \times 3 = 120,960 \text{ lbs.}$$

One-fifth of this should be taken as a safe load,

$$= 24,192 \text{ lbs.} = 10 \text{ tons } 16 \text{ cwt.}$$

which is the safe load required.

To find the sectional area of a cylindrical beam
square its diameter and multiply by ·7854.

The effect of fracture of a member of a scaffold
depends upon its cause and upon the importance of the
member destroyed.

If the fracture is caused by a live load, say a heavy
stone being suddenly placed over a putlog, it is probable
that the suspending rope, if still attached, would prevent
more damage being done. If the fracture arose from an
increasing dead load, say a stack of bricks being gradually built up by labourers, the mass would probably
tear its way through all obstructions. Nevertheless, the
entire scaffold, if well braced and strutted, should not
come down, the damage remaining local.

The result of fracture of a standard under direct
ushing would be somewhat different, as, providing that
ie scaffold is rigid, the greater strain thrown upon the

ledgers, due to the increased distance between supports, would probably cause them to fracture.   In this case the damage would probably still remain local.   If, owing to the fracture, the effect of the bracing were lost, the whole scaffold would probably fail, as shown in the chapter on Stability.

It should be noted that the ledgers, together with the putlogs when fixed at both ends, apart from carrying the loads, have an important effect upon the standards, as, when securely connected, they divide the uprights into a series of short posts, thus dispelling any likelihood of failing by flexure.

154

THE

THE safety of
stability and
use of certain

The stay should be a wooden shaft with an iron clip. By clipping the rung as shown in fig. 139 they do not meet the

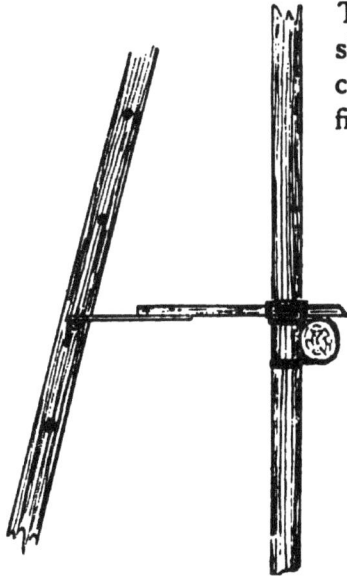

FIG. 139

workman's hands and feet when climbing. The same effect is gained when the top of the ladder rises considerably above the point of rest, by staying as shown on fig. 140.

Ladders should have a level footing and be firmly tied to the point of rest.

Working platforms should be fitted with guard rails along the outside and at the

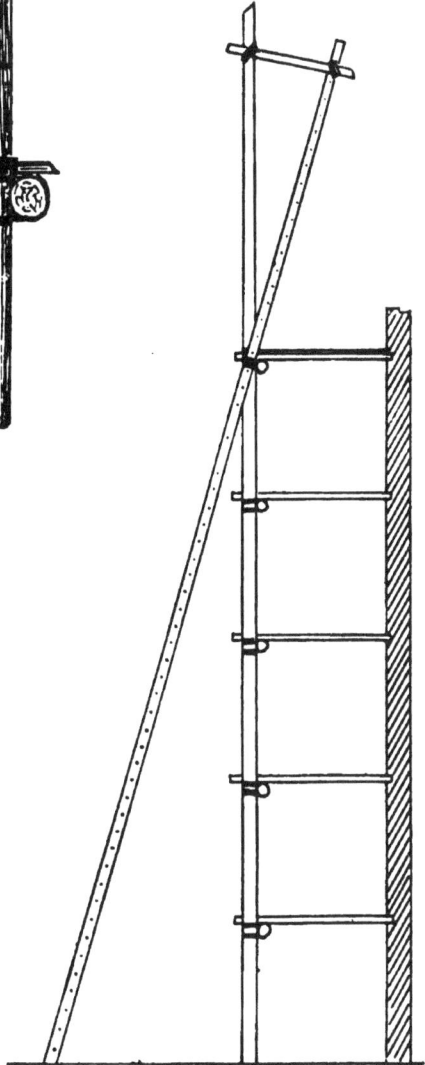

FIG. 140

ends, at a height
They may be tem
the workmen and
necessary to do so i

If a well is le
which to hoist mat
with rails as for the
  A well hole, if l
be fitted with a hin
as required.
  Boards on edge
ends of working p
platform at least 7
board being used

Edge boards are usually nailed to the standards. On exposed situations it is better to tie them, as the wind, continually acting on their surface, will in time draw the nails.

Platform boards, when lapping, frequently lose their place, being kicked by the workmen during their progress about the scaffold. When this happens the boards assume the position shown on fig. 141, and what is known as a trap is formed. The danger of a trap is shown on fig. 142.

Platform boards to be safe from tilting, should not project more than 6 inches beyond the putlogs. At this distance the weight of the workman is most over the putlog, and even if he stood on the extreme edge, experiments have shown that his weight

FIG. 143

is more than counterbalanced by the weight and length of an ordinary board.

Where scaffold boards are used as a means of communication between one part of the scaffold and another they should be laid in pairs, so as to form 'runs' at least 18 inches wide. To prevent unequal sagging they should be strapped on the under side. It would be better to have properly constructed gangways and most decidedly safer.

'Bridging runs' for barrows are usually three boards wide. Five boards wide is better, and, as previously

158

shown, **they**
ging.
    Centering
from a solid

Only the scaffolder or his assistant should be allowed to erect, alter or adapt the scaffolding for its different purposes. Many accidents, again, occur owing to the scaffolding having been altered during a temporary absence of the mechanic, and the reconstruction not having been made safe by his return. This most frequently happens when the scaffolding is not under the charge of one responsible person.

No working platform should be used by the mechanic until its construction is complete. Sufficient plant should be on the job to enable this to be done without disturbing the platform already in use.

Scaffolds should not be heavily loaded. Apart from the risk of the timbers failing, the weight, in the case of the bricklayers' scaffold, has a bad effect upon the new work.

Fan guards, as shown on fig. 145, are usually erected in urban districts to safeguard the public from falling material. There is no

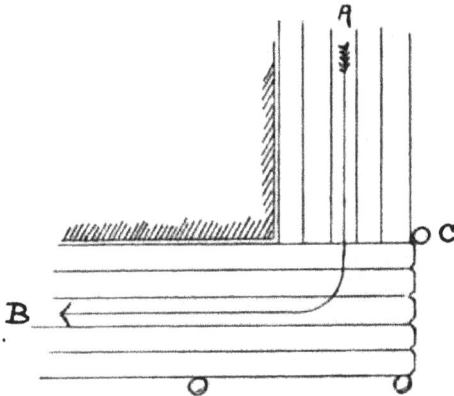

FIG. 146

reason why, for the safety of the workmen, they should not be always fixed.

Due care should be exercised by the workmen themselves, and observance made to the unwritten rules of experience.

The following instance is given as an illustration of what is meant.

A scaffolder
on fig 146 to po
right shoulder, a
against the stan
scaffold.   If he
he would have f.
would not have

CHAPTER XI

LEGAL MATTERS AFFECTING SCAFFOLDING. BUILDING BYE-LAWS

THE following regulations for the erection of scaffolding within the City boundaries have been issued by the Corporation of London, and may be taken as typical of their kind.

CORPORATION OF LONDON

REGULATIONS FOR SCAFFOLDS

APPLICATIONS

Each application for a scaffold is to be entered in a book, with headings for the following information :—

Name of street or place, and number of house.

Nature of work to be executed.

Area of ground level of new premises to be built, or old premises largely altered.

Number of storeys, including ground floor, if new premises are to be built or old premises altered.

Length of scaffolding needed.

Time for which license is requested.

Name and address of Owner.

Do.   do.   Architect.

Do.   do.   Builder.

Date of Application.

Signature of Applicant.

REGULATIONS

The Inspector of Pavements is to report in the application book the time he thinks needful for the scaffold to be licensed ; the

M

license is then to be made out, and the conditions entered in the book by the Engineer's Clerk.

If there is disagreement between the applicant and the Inspector, as to the time needed, the Engineer will decide.

No scaffold is to project beyond the foot-way pavement where it is narrow, nor more than 6 feet where it is wide enough to admit of such projection; any deviation on account of special reasons is to be stated upon the license.

No scaffold is to be enclosed so as to prevent passengers passing under it.

The lower stages of scaffolds are to be close or doubly planked; each stage to have fan and edge boards, and such other precautions to be taken as the Inspector of Pavements requires, to prevent dirt or wet falling upon the public, or for the public safety.

No materials are to be deposited below any scaffold.

Where practicable or needed, a boarded platform, 4 feet wide, and as much wider as may be necessary for the traffic, with stout post rails, and wheel kerbs on the outside of it, are to be constructed outside the scaffold, as the Inspector may direct.

Where it is necessary in the public interest, applicants shall form a gantry, stage, or bridge over the public-way, if required, so as to allow the foot passengers to pass beneath it. The gantry is to be double planked, and so constructed as to prevent dust, rubbish, or water falling upon the foot passengers, and the licensee shall keep the public-way beneath it clean to the satisfaction of the Inspector.

Scaffolds are to be watched and lighted at night.

All fire hydrants must be left unenclosed in recesses formed of such size and in such manner as may enable the hydrant to be easily got at and used.

Public lamps are not to be enclosed without the permission of the Engineer. When such enclosure is permitted, the applicant shall put a lamp or lamps temporarily outside the scaffold, so that the public-way may be properly lighted.

The licensee shall undertake to employ and pay the Contractors to the Corporation to make good the pavements, lamps, and all works disturbed, to the satisfaction of the Engineer.

Licenses are not allowed to be transferred.

FACTORY AND WORKSHOP ACT, 1901

Section 105 of the above Act, so far as it relates to buildings, reads as follows :—

BUILDINGS

105.—(1). *The provisions of this Act with respect to —*

(1) *power to make orders as to dangerous machines (section 17) ;*

(2) *accidents (sections 19-22) ;*

(3) *regulations for dangerous trades (sections 79-86) ;*

(4) *powers of inspection (section 119) ; and*

(5) *fines in case of death or injury (section 136)*

shall have effect as if any premises on which machinery worked by steam, water, or other mechanical power, is temporarily used for the purpose of the construction of a building or any structural work in connection with a building were included in the word ' factory,' and the purpose for which the machinery is used were a manufacturing process, and as if the person who, by himself his agents, or workmen, temporarily uses any such machinery for the before-mentioned purpose were the occupier of the said premises ; and for the purpose of the enforcement of those provisions the person so using any such machinery shall be deemea to be the occupier of a factory.

(2). The provisions of this Act with respect to notice of accidents, and the formal investigation of accidents, shall have effect as if any building which exceeds 30 feet in height, and which is being constructed or repaired by means of a scaffolding . . . . were included in the word ' factory,' and as if . . . . the employer of the persons engaged in the construction or repair . . . . were the occupier of a factory.

It will be noticed that the provisions of the Act are more stringent for buildings which are being constructed or repaired· by machinery, and that these buildings come within the provisions of the Act whether or not they exceed the limit of 30 feet.

The provisions of the Act as mentioned in the begin-

164

ning of this section
abstract, issued fro

Form 57.*
January 1902.

FACTORY A

Abstract o

BUILDINGS IN C

H.M. INSPECTOR
To whom Con
Notices sh

H.M. SUPERINTEN
SPECTOR OF F.

H.M. CHIEF INSP
FACTORIES

CERTIFYING SURG

The provisions stated below apply also to any private line or siding used in connection with a building in course of construction or repair as above.

Dangerous Machinery or Plant. 1.—If any part of the ways, works, machinery, or plant (including a steam boiler) is in such condition that it cannot be used without danger to life or limb, a Court of Summary Jurisdiction may, on complaint of an Inspector, make an order prohibiting it from being used, absolutely or until it is duly repaired or altered.

Dangerous Processes. 2.--If any machinery, plant, process, or description of m nual labour is dangerous or injurious to health, or dangerous to life or limb, regulations may be made by the Secretary of State.

Steam boilers. 3.—Every steam boiler must (a) be maintained in proper condition, and (b) have a proper safety-valve, steam-gauge, and water-gauge, all maintained in proper condition, and (c) be thoroughly examined by a competent person every 14 months. A signed report of the result of the examination must be entered within 14 days in a Register to be kept for the purpose in the premises (Form 73*).

Accidents. 4.—When there occurs in the premises any accident which causes to a person employed therein such injury as to prevent him on any one of the three working days next after the occurrence of the accident from being employed for five hours on his ordinary work, written Notice (Form 43*) must be sent forthwith to H.M. Inspector for the district.

5.—Every such accident must also be entered in a Register to be kept for the purpose in the premises (Form 73*).

6.—If the accident is fatal, or is produced by machinery moved by power, or by a vat or pan containing hot liquid, or by explosion, or by escape of gas or steam, written Notice (Form 43*) must also be sent forthwith to the Certifying Surgeon for the district.

Returns. 7.—If so required by the Secretary of State, a return of the persons employed must be sent to H.M. Chief Inspector of Factories at such times and with such particulars as may be directed.

Powers of Inspectors. 8.—H.M. Inspectors have power to inspect every part of the premises by day or by night. They may require the production of registers, certificates, and other papers. They may examine any person found in the premises either alone or in the presence of any other person as they think fit, and may require him to sign a declaration of the truth of the matters about which he is examined. They may also exercise such other powers as may be necessary for carrying the Act into effect. Every person obstructing an Inspector, or refusing to answer his questions, is liable to a penalty.

The limiting height of 30 feet has been inserted for the reason, apparently, that it was not considered desirable to bring those minor accidents which might reasonably be expected to occur on the smaller buildings within the provisions of the Act.

As stated in paragraph 2 of the Abstract, the Secretary of State is empowered to make regulations for any description of manual labour that is dangerous or injurious to health, or dangerous to life or limb.

No regulations for the building trade have as yet been made, but early in 1902 a memorandum on building accidents and their prevention was issued generally to master builders, and the following suggestions embodied therein.

1. *All working platforms above the height of 10 feet, taken from the adjacent ground level, should, before employment takes place thereon, be provided throughout their entire length on the outside and at the ends,*

    (*a*) *with a guard rail fixed at a height of 3 feet 6 inches above the scaffold boards. Openings may be left for workmen to land from the ladders, and for the landing of material.*

    (*b*) *with boards fixed so that their bottom edges are resting on or abutting to the scaffold boards. The boards so fixed should rise above the working platform not less than 7 inches. Openings may be left for the landing of the workmen from the ladders.*

2. *All 'runs' or similar means of communication between different portions of a scaffold or building should be not less than 18 inches wide. If composed of two or more boards they should be fastened together in such a manner as to prevent unequal sagging.*

3. *Scaffold boards forming part of a working platform should be supported at each end by a putlog, and should not project more than 6 inches beyond it unless lapped by another board, which should rest partly on or over the same putlog and partly upon putlogs other than those upon which the supported board rests.*

   In such cases where the scaffold boards rest upon brackets, the foregoing suggestion should read as if the word bracket replaced the word putlog.

   N.B.—Experiments have shown that a board with not more than a 6-inch projection over a putlog can be considered safe from trapping or tilting.

4. *All supports to centering should be carried from a solid foundation.*

5. *In places where the scaffolding has been sublet to a Contractor, the employer should satisfy himself, before allowing work to proceed thereon, that the foregoing suggestions have been complied with, and that the material used in the construction of the scaffold is sound.*

In many cases these suggestions are already carried out by builders, and even in some cases improved upon. For instance, edge boards are often supplemented by additional boards rising to two or three feet in height.

Their use for the prevention of material falling is undoubtedly beneficial. Unfortunately edge boards cannot be fixed on the side of the platform nearest to the building as they would interfere with the free use of the workmen's tools.

The fixing of guard rails should be considered an absolute necessity to safeguard a working platform. The danger of workmen falling is a real one, and does not always arise from loss of nerve: tripping against material, platform boards, and stepping backwards to view progress of work, accounting for many accidents. The rails have often to be removed for the landing of material, but if this is not absolutely necessary they should be allowed to remain.

The second suggestion is obviously meant to apply to bridging runs only.

The limit of a 6-in. projection to scaffold boards, apart from whether the board would tilt or not, is a safeguard in that it prevents the workmen from standing on the boards outside of the end putlogs. This is especially necessary at the ends of the platforms, if it is remembered that when on a scaffold it is not easy to notice the exact position of the putlogs.

WORKMEN'S COMPENSATION ACT, 1897

LIABILITY OF CERTAIN EMPLOYERS TO WORKMEN FOR INJURIES

Sec. 1.—(1) *If in any employment to which this Act applies personal injury by accident arising out of and in the course of the employment is caused to a workman, his employer shall, subject as hereinafter mentioned, be liable* ˙˙ *pay compensation in accordance with the First Schedule* ˙*is Act.*

(2) *Provided that :*

    (*a*) *The employer shall not be liable under this Act in respect of any injury which does not disable the workman for a period of at least two weeks from earning full wages at the work at which he was employed ;*

    (*b*) *When the injury was caused by the personal negligence or wilful act of the employer, or of some person for whose act or default the employer is responsible, nothing in this Act shall affect any civil liability of the employer, but in that case the workman may, at his option, either claim compensation under this Act, or take the same proceedings as were open to him before the commencement of this Act ; but the employer shall not be liable to pay compensation for injury to a workman by accident arising out of and in the course of the employment both independently of and also under this Act, and shall not be liable to any proceedings independently of this Act, except in case of such personal negligence or wilful act as aforesaid ;*

    (*c*) *If it is proved that the injury to a workman is attributable to the serious and wilful misconduct of that workman, any compensation claimed in respect of that injury shall be disallowed.*

(3) *If any question arises in any proceedings under this Act as to the liability to pay compensation under th Act (including any question as to whether the employm is one to which this Act applies), or as to the amou*

duration *of compensa*
not settled *by agreeme*
the First *Schedule to*
accordance *with the*

(4) *If,* **within** 1
limited *for* **taking** p
recover **damages** ina
caused by **any acciden**
that the **injury is one**
in such **action, but tha**
compensation **under** t
shall be **dismissed ;**
tried shall, **if the plai**
such **compensation,** a
such **compensation** al
have been **caused by**
instead of **proceeding**

tractor of any work, and the undertakers would, if such work were executed by workmen immediately employed by them, be liable to pay compensation under this Act to those workmen in respect of any accident arising out of and in the course of their employment, the undertakers shall be liable to pay to any workman employed in the execution of the work any compensation which is payable to the workman (whether under this Act or in respect of personal negligence or wilful act independently of this Act) by such contractor, or would be so payable if such contractor were an employer to whom this Act applies.

Provided that the undertakers shall be entitled to be indemnified by any other person who would have been liable independently of this section.

This section shall not apply to any contract with any person for the execution by or under such contractor of any work which is merely ancillary or incidental to, and is no part of, or process in, the trade or business carried on by such undertakers respectively.

APPLICATION OF ACT AND DEFINITIONS

Sec. 7.—*This Act shall apply to . . . employment by the undertakers as hereinafter defined on, in, or about any building which exceeds 30 feet in height, and is either being constructed or repaired by means of a scaffolding, or being demolished, or on which machinery driven by steam, water, or other mechanical power is being used for the purpose of the construction, repair, or demolition thereof.*

It will be noticed that this Act applies in like manner as the Factory Acts, except that, in addition, buildings under demolition, if over 30 feet in heir if mechanical power is being used in the demol' the building, are included. This has had the

greater care being take
to the extent of erecting

In this Act 'Underta
means the person under
or demolition.

'Employer' includes
or incorporate, and the
a deceased employer.

'Workman' includes
in any employment to w
by way of manual labou
agreement is one of ser
wise, and is expressed o
Any reference to a wo
shall, where the workma
to his legal personal repr
or other person to whom

The following cases

as completed until the scaffolding had been removed. This view was upheld by the Court of Appeal.[1]

Repair.—'Repair ' has been held to include whitewashing, stopping, and dubbling a ceiling in a building over 30 feet in height.[2]

Employment on, in, or about a Building.—A bricklayer employed on a new wing to a dwelling-house was injured while at work. There was a cesspool into which the house was drained, situated at a considerable distance from the house, but connected with it by a drain. Close to this cesspool there was a heap of bricks, to be used in repairing it. The bricklayer was chipping off a piece of brick from one of the heap, when a piece flew into his eye and injured it. The County Court judge awarded the bricklayer compensation ; and the employers appealed, contending that, although the main building was over 30 feet in height, and scaffolding was being used on the new wing, there was no evidence to show that the bricklayer at the time of the accident was working on, in, or about a building which exceeded 30 feet in height, and which was either being constructed or repaired by means of a scaffolding. The Court of Appeal held that there was ample evidence on which the County Court judge could come to the conclusion at which he arrived, and dismissed the appeal with costs.[3]

A workman found he wanted to use a pickaxe. There being none on the premises, he went to fetch one from the builder's yard, some distance away. While going he slipped and injured himself. He claimed

[1] Frid v. Fenton, Court of Appeal, 1900.
[2] Dredge v. Conway Jones & Co., Court of Appeal, 1901.
[3] Harrison v. Gutherie & Son, Court of Appeal, 1902.

has been fixed by these Acts to det(
ing exceeds 30 feet in height.

Measurements for this purpose (
the street level, adjacent ground lev
or ground floor, or from the top or l
footings laid—to the highest po
reached if unfinished, or to the rid
completed.

It is apparent that to obtain u
under the Acts, certain fixed poi
measure are necessary. They sho
occur without variation on all bu
should not vary in position on an
levels are not always to be foun
Ground levels vary, often owing to
excavations and other preliminari
construction. The top of the foc
not the lowest portion of the bu
structure reaches to a height of 3(
not always completed. This leaves
footings as being the lowest fixed
measurements can be taken. This

was evidence that, at the time of the accident, the building was more than 30 feet in height, whether measured from the top or bottom of the footings, that the footings had been filled in, and that the building had not reached the stage at which more than the footings had been filled in.   There was an appeal, but the decision was confirmed.[1]

A firm of builders were demolishing a house, and at the time of the accident it was only 11 feet in height, with the exception of the party-wall, which exceeded the 30 feet limit.   It was held that, as this wall remained standing, it was evidence upon which to find that the building exceeded the 30 feet in height.[2]

A workman was injured by the demolition of a building of less than 30 feet in height, but which was connected internally to an adjacent building which exceeded that height, and which belonged to the same owner.   The Court of Appeal held that the building demolished did not exceed 30 feet in height.[3]

Scaffolding.—Scaffolding includes an internal staging of planks resting partly upon a ladder step and otherwise upon a member of a roof truss.[4]   Scaffolding means also a platform of boards on ledgers tied to iron columns and supported by trestles.[5]

In the following case the walls of a refuse destructor consisted of open arches.   It was resolved to fill up these arches with panels.   A workman, while fixing the frames to hold these panels, met with an accident. The County Court judge held, in the first place, that the

---

[1] McGrath v. McNeil & Son, Court of Appeal, 1901.
[2] Knight v. Cubitt & Co., Court of Appeal, 1900.
[3] Rixsome v. Pritchard and Another, Court of Appeal, 1900.
[4] Reddy v. Broderick, 1901.
[5] Hoddinott v. Newton, Chambers & Co., House of Lords, 1901.

work upon which this workman was engaged was not
'construction,' and, secondly, that a plank supported by
two piles of slag was not 'scaffolding.' He therefore
refused to award him compensation   The workman appealed ; and the Court of Appeal sent the case back to
the County Court judge for him to state his view of the
facts as to what was actually the work in question, and
his reasons for holding that it was not 'construction.' The
Court held that the County Court judge was wrong in
holding that the plank in question was not 'a scaffolding.'[1]

In another case a workman, employed by a builder,
who was executing repairs to the roof of a house exceeding 30 feet in height, was carrying slates up a ladder,
when it slipped and he fell and was injured.   The ladder
in question was placed against the house to enable
workmen to get on to the roof from the street, one end
of the ladder resting on the ground and the other end
leaning against the parapet at the top of the house.
There was no ladder, crawling board, or other contrivance on the roof.   The County Court judge held that
the ladder was not a 'scaffolding,' within the meaning
of the Act, and refused to award the workman compensation.   He appealed : and the Court of Appeal declined
to hold that the County Court judge had misdirected

tracted to construct a house, sublet the plastering to a sub-contractor, to whom they supplied all materials. One of the sub-contractor's workmen, having met with an injury while at work, claimed compensation from the builders. They thereupon brought in the sub-contractor as a third party, claiming that he was liable to indemnify them. The County Court judge made an award in favour of the workman, but declared that the builders were not entitled to be indemnified by the sub-contractor. The builders appealed from this part of the award. The sub-contractor did not appear on the hearing of the appeal. The Court of Appeal allowed the appeal, with costs, holding that the sub-contractor, having undertaken a substantial part of the construction of the building, was himself an ' undertaker,' and was therefore liable to indemnify the builders.[1]

A firm of builders contracted to erect a building. They arranged with a sub-contractor to do the slating work. A labourer in the employment of this sub-contractor was killed by an accident, and compensation was awarded to his widow against the contractors, who claimed to be indemnified by the sub-contractor. The County Court judge held the contractors entitled to be so indemnified ; but this decision was reversed by the Court of Appeal. On appeal to the House of Lords that House reversed the decision of the Court of Appeal.[2]

[1] Wagstaff *v.* Perks & Son ; Firth, third party, Court of Appeal, 1902.
[2] Cooper & Crane *v.* Wright, House of Lords, 1902.

APPENDIX

WEIGHT OF MATERIAL

	lbs. per cubic foot
ASPHALTE (gritted)	156·00
Brick (common) from	97·31
,, ,, to	125·00
,, (red)	135·50
,, (pale red)	130·31
,, (Common London Stock) . . .	115·00
,, paving (English clinker)	103·31
,, ,, (Dutch ,,)	92·62
Brickwork in mortar, about	110·00
Cement (Roman) and sand (equal parts) . . .	113·56
,, alone (cast)	100·00
Chalk from	125·00
,, to	166·00
,, (Dorking)	116·81
Clay (common)	119·93
,, (with gravel)	160·00
Concrete (lime)	130·00
,, (cement)	136·00
Earth (common) from	95·00
,, ,, to	124·00
,, (loamy)	126·00
., (rammed)	99·00
,, (loose or sandy)	95·00
Firestone	112·00

Iron (bar) ... ...

,,	,,	to	.	.	.	.
,,	(hammered)		.	.	.	.
,,	(not hammered)		.	.	.
,,	(cast) from		.	.	.	.
,,	,, to	.	.	.	.	.
Lead (milled)	.	.	.	.	.	.
,, (cast)	.	.	.	.	.
Lime (quick)	.	.	.	.	.
Marble from	.	.	.	.
,, to	.	.	.	.	.
Mortar (New)	.	.	.	.
,, (of river sand 3 parts, of lime
,, well beaten together	.
,, (of pit sand) .	.	.
,, well beaten together	.
,, (common of chalk, lime and s
,, (lime, sand, and hair for plast
Sand (pure quartz)	.	.	.
,, (river)	.	.	.	.
,, (River Thames best)	.
,, (pit, clean but coarse)	.
,, (pit, but fine grained)	.
,, (road grit)	.	.	.

,, Thames inferior)

	lbs. per cubic foot
Slate, Welsh rag	172·00
„ Cornwall (greyish blue)	157·00
Snow from	8·00
„ to	14·00
Shingle	95·00
Steel from	486·85
„ to	490·00
Stone, Bath (roe stone)	155·87
„ „	123·43
„ blue lias (limestone)	154·18
„ Bramley Fall (sandstone)	156·62
„ Bristol	156·87
„ Caen	131·75
„ Clitheroe (limestone)	167·87
„ Derbyshire (red friable sandstone)	146·62
„ Dundee	158·12
„ Hilton (sandstone)	136·06
„ Kentish rag	166·00
„ Ketton (roe stone)	155·87
„ Kincardine (sandstone)	153·00
„ Penarth (limestone)	165·81
„ Portland (roe stone)	153·81
„ „	151·43
„ Purbeck	167·50
„ Woodstock (flagstone)	163·37
„ Yorkshire (paving)	156·68
Tile (common plain)	116·15
Water (sea)	64·18
„ (rain)	62·50
Wood (ash)	50·00
„ (birch)	45·00
„ (beech)	45·00
„ (elm, common, dry)	34·00
„ (fir, Memel dry)	34·00
„ („ Norway spruce)	32·00
„ („ Riga dry)	29·12
„ („ Scotch dry)	26·8

Zinc

INDEX

Standard Books for
SURVEYORS, ARCHITECTS,
BUILDERS, CRAFTSMEN, ETC.

PUBLISHED AND SOLD BY

B. T. BATSFORD, 94 High Holborn,

LONDON.

SIXTH EDITION (40TH THOUSAND), REVISED AND GREATLY ENLARGED.

BUILDING CONSTRUCTION AND DRAWING. A
TEXT-BOOK ON THE PRINCIPLES AND PRACTICE OF CONSTRUCTION
specially adapted for Students in Science and Technical Schools. By
CHARLES F. MITCHELL, Lecturer on Building Construction at the
Polytechnic Institute, London. FIRST STAGE, OR ELEMENTARY
COURSE. 400 pp. of Text, with 1,000 Illustrations, fully dimensioned.
8vo. cloth. 3s.

"AN EXCELLENT AND TRUSTWORTHY LITTLE TREATISE, PREPARED AND ILLUSTRATED
IN A VERY THOROUGH AND PRACTICAL SPIRIT."—THE BUILDER.

"It seems to have most of the advantages of Vols. 1 and 2 of Rivington's 'Building
Construction,' with the additional ones of cheapness and conciseness, and appears to be
thoroughly practical."—MR. J. T. HURST, Author of the "SURVEYOR'S HANDBOOK."

"A model of clearness and compression, well written and admirably illustrated,
and ought to be in the hands of every student of building construction."—THE BUILDER.

FOURTH EDITION (18TH THOUSAND), REVISED AND GREATLY ENLARGED.

BUILDING CONSTRUCTION. Advanced and
Honours Courses. By CHARLES F. MITCHELL. For the use
of Students preparing for the Examinations of the Science and Art
Department, the Royal Institute of British Architects, the Surveyors'
Institution, the City Guilds, &c. 700 pp. of Text, with 660 Illustrations,
fully dimensioned. Crown 8vo. cloth, 5s. 6d.

"Mr. Mitchell's two books form unquestionably the best guide to all the mechanical
part of architecture which any student can obtain at the present moment. In fact, so far
as it is possible for anyone to compile a satisfactory treatise on building construction,
Mr. Mitchell has performed the task as well as it can be performed."—THE BUILDER.

BRICKWORK AND MASONRY. A Practical Text-Book
for Students and those engaged in the Design and Execution of Structures in Brick and Stone. By CHARLES F. MITCHELL and GEORGE A.
MITCHELL. Being a thoroughly revised and remodelled edition of the
chapters on these subjects from the Authors' "Elementary" and "Ad-
vanced Building Construction," with special additional chapters and new
illustrations. 300 pp., with 600 illustrations. Crown 8vo. cloth, 5s.

TREATISE ON SHO

and generally dealin
By C. H. STOCK, Architect
revised by F. R. FARROW,
cloth, 4s. 6d.

" The treatise is a valuable addit
builder, and we heartily recommend it t

" Mr. Stock has supplied a manifest
surveying, and there is no doubt that his

DANGEROUS STRUC

Men.   By GEO. H. BLAGRO

" We recommend this book to all
ARCHITECT.

STRESSES AND THR

tural Students.   By G. A. T.
revised, containing new chap
Steel Joist and of a Steel
Rectangular Beams, Arches, .
and Folding Plates.   Crown

MODERN PRACTICAL JOINERY. A Guide to the Preparation of all kinds of House Joinery, Bank, Office, Church, Museum, and Shop-fittings, Air-tight Cases, and Shaped Work, with a full description of Hand-tools and their uses, Workshop Practice, Fittings and Appliances, also Directions for Fixing, the Setting-out of Rods, Reading of Plans, and Preparation of Working Drawings, Notes on Timber, and a Glossary of Terms, &c. By GEORGE ELLIS, Instructor in Joinery at the Trades Training Schools of the Worshipful Company of Carpenters. Containing 380 pp., with 1,000 Practical Illustrations. Large 8vo. cloth, 12s. 6d. net.

"In this excellent work the mature fruits of the first-hand practical experience of an exceptionally skilful and intelligent craftsman are given. It is a credit to the author's talent and industry, and is likely to remain an enduring monument to British craftsmanship. As a standard work it will doubtless be adopted and esteemed by the architect, builder, and the aspiring workman."—BUILDING WORLD.

HOW TO ESTIMATE: or the Analysis of Builders' Prices. A complete Guide to the Practice of Estimating, and a Reference Book of Building Prices. By JOHN T. REA, F.S.I., Surveyor, War Department. With typical examples in each trade, and a large amount of useful information for the guidance of Estimators, including thousands of prices. Second Edition, revised to date, and considerably extended, large crown 8vo. cloth, 7s. 6d. net.

This work deals with the principles and practice of estimating in a thoroughly practical and comprehensive manner, and is the outcome of eighteen years' experience in the personal supervision of large contracts.

It is applicable for pricing in any part of the country, and is adaptable to every class of building and circumstance.

Each chapter deals with a trade, and is divided into three main divisions :

(1) Memoranda, (2) Prices, and (3) Analysis of Materials and Labour.

THE BOOK IS EXCELLENT IN PLAN, THOROUGH IN EXECUTION, CLEAR IN EXPOSITION, AND WILL BE A BOON ALIKE TO THE RAW STUDENT AND TO THE EXPERIENCED ESTIMATOR. FOR THE FORMER IT WILL BE AN INVALUABLE INSTRUCTOR; FOR THE LATTER A TRUSTWORTHY REMEMBRANCER AND AN INDISPENSABLE WORK OF REFERENCE.—THE BUILDING WORLD.

BUILDING SPECIFICATIONS, for the use of Architects, Surveyors, Builders, &c. Comprising the Complete Specification of a Large House, consisting of 714 numbered clauses ; also numerous clauses relating to special Classes of Buildings, as Warehouses, Shop Fronts, Public Baths, Schools, Churches, Public-Houses, &c., &c., and Practical Notes on all Trades and Sections. By JOHN LEANING, F.S.I., Author of "Quantity Surveying," &c. 650 pp., with 150 Illustrations. Large 8vo. cloth, 18s. net.

"Cannot but prove to be of the greatest assistance to the specification writer, whether architect or quantity surveyor, and we congratulate the author on the admirable manner in which he has dealt with the subject."—THE BUILDER'S JOURNAL.

"A very valuable book. It must become a standard work."—THE BRITISH ARCHITECT.

THE LONDON BUILDING ACTS, 1894-98. A Text-

Book on the Law relating to Building in the Metropolis. Containing the Acts, printed *in extenso*, with a full Abstract giving all the Sections of the 1894 Act which relate to building, set out in Tabular Form for easy reference, together with the unrepealed Sections of all other Acts affecting building and the latest Bye-Laws and Regulations. Third Edition, thoroughly revised by BANISTER F. FLETCHER, A.R.I.B.A., F.S.I., and H. PHILLIPS FLETCHER, F.S.I., Barrister-at-Law, with abstracts of the latest decisions and cases. With 23 Coloured Plates, showing the thickness of walls, plans of chimneys, &c. Crown 8vo. cloth, 6s. 6d.

"It is the Law of Building for London in one volume."—ARCHITECT.

"The Abstract of the portion of the Act relating to building is very useful as a finger-post to the Sections in which the detailed regulations in regard to various operations of building are to be looked for—an assistance the more desirable from the fact that the Act is by no means well or systematically arranged.'—THE BUILDER.

"Illustrated by a series of invaluable coloured plates, showing clearly the meaning of the various clauses as regards construction."—THE SURVEYOR.

CONDITIONS OF CONTRACT. A Work dealing with

Conditions of Contracts and with Agreements as applied to Building Works, and with the Law generally in its relation to various matters coming within the scope of the Architectural Profession. By FRANK W. MACEY, Architect, Author of "Specifications in Detail." Revised, as to the strictly legal matter, by B. J. LEVERSON, Barrister-at-Law. Royal 8vo. cloth, 15s. net.

ESTIMATING: A Method of Pricing Builders' Quantities for Competitive Work. By GEORGE STEPHENSON.

Showing how to price, *without the use of a Price Book*, the Estimates of the work to be done in the various Trades throughout a large Villa Residence. Fifth Edition, the Prices carefully revised. Crown 8vo. cloth, 4s. 6d. net.

"The author, evidently a man who has had experience, enables everyone to enter, as it were, into a builder's office and see how schedules are made out. The novice will find a good many 'wrinkles' in the book."—ARCHITECT.

REPAIRS : How to Measure and Value Them. A

Handbook for the use of Builders, Decorators, &c. By the Author of "Estimating." Third Edition, the prices carefully revised. Crown 8vo. cloth, 3s. 6d.

"'Repairs' is a very serviceable handbook on the subject. A good specification for repairs is given by the author, and then he proceeds, from the top floor downwards, to show how to value the items, by a method of framing the estimate in the measuring book. The *modus operandi* is simple and soon learnt."—THE BUILDING NEWS.